Energy Science, Engineering and Technology

SUSTAINABLE PETROLEUM OPERATIONS

Energy Science, Engineering and Technology

Additional books in this series can be found on Nova's website under the Series tab.

Additional E-books in this series can be found on Nova's website under the E-book tab.

Energy Science, Engineering and Technology

SUSTAINABLE PETROLEUM OPERATIONS

RAFIQ ISLAM
EDITOR

Nova Science Publishers, Inc.
New York

For permission to use material from this book please contact us:
Telephone 631-231-7269; Fax 631-231-8175
Web Site: http://www.novapublishers.com

Library of Congress Cataloging-in-Publication Data

Sustainable petroleum operations / editor, Rafiq Islam.
p. cm.
Includes index.
ISBN 978-1-61324-157-8 (hardcover)
1. Petroleum engineering. 2. Oil spills--Prevention. 3. Sustainable engineering. 4. Petroleum engineering--Environmental aspects. I. Islam, Rafiqul, 1959-
TN870.S835 2011
333.8'232--dc22

2011008482

Published by Nova Science Publishers, Inc. † New York

CONTENTS

PREFACE

This new book presents and discusses current research in the study of sustainable petroleum operations. Topics discussed include greening of petroleum operations; a novel environmental risk management model for petroleum operations; an approach for conflict resolution in oil and gas operations; technological analysis and quantitative assessment of oil and gas development and air pollution and carbon assessment framework for oil and gas developments.

Chapter 1 - The evolution of human civilization is synonymous with how it meets its energy needs. Few would dispute the human race has become progressively more civilized with time. Yet, for the first time in human history, an energy crisis has seized the entire globe and the very sustainability of this civilization itself has suddenly come into question. If there is any truth to the claim that Humanity has actually progressed as a species, it must exhibit, as part of its basis, some evidence that overall efficiency in energy consumption has improved. In terms of energy consumption, this would mean that less energy is required per capita to sustain life today than, say, 50 years earlier. Unfortunately, exactly the opposite has happened. The author used to know that resources were infinite, and human needs finite. After all, it takes relatively little to sustain an individual human life. Things have changed, however, and today they are told, repeatedly: resources are finite, human needs are infinite.

Chapter 2 – Oil and gas are considered the lifeblood of modern economic society. However, the exploration and development of oil and gas have recently been found to be environmentally unsustainable. Each phase of hydrocarbon extraction, especially in offshore operations, generates numerous gaseous, liquid, and solid wastes. This paper examines the present status of

Offshore Oil and Gas Operations (OOGOs) with respect to waste generation, resource abuse, and toxic compounds usages in its lifecycle. This research analyzes the sustainability of offshore oil and gas operations using the concept of 'green supply chain'. It studies environmental dimensions, company policy, operations, and technology in the context of offshore oil and gas development in the Scotian shelf of Canada. To achieve sustainability in OOGOs, an 'Olympic' green supply chain model has been applied to achieve zero emissions, zero waste of resources, zero waste in activities, zero use of toxics, and zero waste in product life-cycle. However, it has been found that there is no technology available to achieve the objectives of the 'Olympic' supply chain model. Finally, this paper discusses the environmental and technological problems of OOGOs and provides guidelines to achieve overall sustainability in oil company activities.

Chapter 3 – Contaminated water is an environmental threat due to its toxicity and potential ecological consequences. Oil and gas operations introduce the enormous amount waste stream in the marine environment. The offshore environment of Eastern Canada is not an exception. Managing this huge volume of water in a coastal domain is indeed a challenge. Fisheries are treated as the valued ecosystem component in the coastal zone of Nova Scotia. The paper examines the different management options in regards to valued fisheries components. It addresses the critical issues in managing the ecological risk generated from produced water to the valued ecosystem component with particular reference to fisheries as important component for the people in the coastal zone of Nova Scotia. An integrated risk management model has been proposed that effectively integrates the possible means to reduce the toxicity and volume and other issues posed by produced water from offshore operations. This model can be applied in any offshore hydrocarbon operations to protect vulnerable valued ecosystem.

Chapter 4 – A persuasive analysis of a complex conflict was carried out regarding oil and gas activities in Cape Breton, Nova Scotia. The introduction sets this particular problem based on literature study, with a number of relevant variable sets discussed. The various stakeholders and their position in oil and gas operations created a number of conflicts. The position of the province is a fair rendition of the view of socio-economic development. Fisheries viewpoint was found very significant. The environmental position with a sharp elaboration of the alliance between environmental and fisheries interests, the probable divisions within the general public were discussed. Analysis and recommendations pull together some of the major axes of conflict and pointed out some of the particular difficulties to be dealt with.

Some models have been proposed to resolute the conflict. The excessively sectoral approach of many agencies, even presumably connected ones, envisaged Integrated Coastal Management (ICM) principles as an approach for resolution.

Chapter 5 – Offshore oil and gas development is a high technology intensive operation. In this paper, an overview of oil and gas development on the Scotian Shelf of Nova Scotia, Canada, and its technological details are examined. For the purpose of risk and environmental impact assessment, the detailed technological phases and scale of oil and gas operations are analysed. The technology phases are categorized into four phases: seismic exploration, exploratory drilling and installation of structures, production, and decommissioning of the project. The technological details and scale of operation involved for each of these phases varies significantly. Each technological phase is critically analysed and it is found that each phase produces/releases numerous wastes, none of which is remedied completely. Besides, each phase of the operation is inherently inefficient and wastefully in energy and mass. This paper also studies, the typical candidate locations and jurisdictional situation for oil and gas project lifecycles. It provides quantitative information of offshore oil and gas development activities in the Scotian Shelf and uses this as a basis for risk and environmental impact assessment. This study will be helpful to engineering and policy makers and can bridge the gap between the two.

Chapter 6 – In order to detect petroleum contaminants in groundwater, it is vital to design and optimize the monitoring network properly and efficiently. So far, the best optimization techniques to solve this type of problems use genetic algorithms. The ability of Genetic Algorithm (GA) methods to identify optimal solutions in groundwater management optimization problems has been widely demonstrated. However, one of the main concerns in applying these methods is identifying suitable values for the GA parameters, especially population size, crossover probability, mutation probability, and usage of elitism. The values selected for these parameters have a significant impact on the objective function solution as well as the time taken to find that solution. Due to interaction of these parameters, the impact of each parameter depends on the values of the other parameters. This paper uses sensitivity analyses as a tool to identify the suitable values for GA parameters for a groundwater monitoring problem. Through these analyses, suitable parameter values are identified that cause the algorithm to perform very well in terms of finding final optimal solution.

Chapter 7 – Sundarbans mangrove ecosystem, the world largest mangrove forest in Bangladesh, is one of the most productive and biologically diverse wetlands in the world. This unique coastal tropical forest is among the most threatened habitats on earth. This paper critically examines the present status of the Sundarbans and assesses the potential threat of oil and gas exploration. Every aspect of the exploration and development of oil and gas is analyzed and the magnitude of impact quantified. To avoid or minimize the identified impacts a mitigation plan and guidelines are addressed. Since there is limited research done regarding these issues, this work can serve as a basis for understanding the effects of oil and gas in an environmentally sensitive area and point the way to further study.

Chapter 8 – Oil and gas operations causes numerous environmental problems. Air pollution is one of the major concerning issue in the whole operations. This paper examines the air pollution causes by the different oil and gas activities. Major air emissions are identified and these are: H_2S, natural gas, carbon monoxide, particulate maters, NO_x, volatile organic carbons, BTEX (Benzene, Toluene, Ethylbenzene, and Xylenes), Polycyclic aromatic hydrocarbons (PAHs), and CO_2. Among all of the pollutants, the CO_2 emission is the highest in quantity. As a result, special attention is given to assess the carbon. A framework is provided to conduct environmental assessment of Carbon as a valued ecosystem compound. A case study was carried out in the offshore oil and gas development on the Scotian Shelf of Nova Scotia, Canada, to justify this proposed framework.

In: Sustainable Petroleum Operations
Editor: Rafiq Islam
ISBN 978-1-61324-157-8

Chapter 1

GREENING OF PETROLEUM OPERATIONS

Arjun B. Chhetri and M. Rafiqul Islam
Faculty of Engineering, Dalhousie University,
Halifax, Nova Scotia, Canada

1. INTRODUCTION

The evolution of human civilization is synonymous with how it meets its energy needs. Few would dispute the human race has become progressively more civilized with time. Yet, for the first time in human history, an energy crisis has seized the entire globe and the very sustainability of this civilization itself has suddenly come into question. If there is any truth to the claim that Humanity has actually progressed as a species, it must exhibit, as part of its basis, some evidence that overall efficiency in energy consumption has improved. In terms of energy consumption, this would mean that less energy is required per capita to sustain life today than, say, 50 years earlier. Unfortunately, exactly the opposite has happened. We used to know that resources were infinite, and human needs finite. After all, it takes relatively little to sustain an individual human life. Things have changed, however, and today we are told, repeatedly: resources are finite, human needs are infinite. What's going on?

Some Nobel Laureates (e.g., Robert Curl) or environmental activists (e.g., David Suzuki) have blamed the entire technology development regime. Others have blamed petroleum operations and the petroleum industry. Of course, in the teeth of increasing oil prices, the petroleum sector becomes a particularly easy target. Then, there is U.S. President George W. Bush, talking about 'oil addiction'. Even his most ardent detractors embrace this comment as some sign of deep thinking. Numerous alternate fuel projects have been launched – but what do they propose? The same inefficient and contaminated process that got us in trouble with fossil fuel! Albert Einstein famously stated, "The thinking that got you into the problem, is not going to get you out." As Enron, once touted as the 'most creative energy management' organization, collapsed, everyone seems more occupied with trying to recoup by using the same (mis-) management scheme that led to its demise.

In this chapter, the mysteries of current energy crisis is unraveled. The root causes of unsustainability in all aspects of petroleum operations are discussed. It is shown how each practice follows a pathway that is inherently implosive. It is further demonstrated that each pathway leads to irreversible damage to the ecosystem and can explain the current state of the Earth. It is shown that fossil fuel consumption is not the culprit, it is rather the practices involved during exploration all the way to refining and processing that are responsible for the current damage to the environment. Discussion is carried out based on two recently developed theories, namely the theory of inherent sustainability and the theory of knowledge-based characterization of energy sources. These theories explain why current practices are inherently inefficient and why new proposals to salvage efficiencies do not have any better chance to remedy the situation. It is recognized that critiquing current practices may be necessary, but it is not sufficient. The second part of the chapter deals with practices that are based on long-term. This can be characterized as the approach of obliquity, which is well-known for curing both long-term and short-term problems. It stands 180 degrees opposite to the conventional band-aid approach, which has prevailed in the Enron-infested decades. This chapter promises greening of every practice of the petroleum industry, from management style to upstream to downstream. In the past, petroleum engineers only focused on drilling, production and transportation, and reservoir engineering. This chapter also deals with management and continues through exploration all the way up to refining and gas processing. In the past, exploration meant increasing the chance of production. Many of these cases ended up costing Humanity much more in the form of environmental damages, which, as this chapter points out, were never part of the management equation. The main objective of this chapter is to lay out the framework of sustainability in the petroleum sectors pertaining to environmental, technological and management aspects. In addition, the framework of truly 'sustainable management' for practically all aspects oil and gas operations will be provided. With the greening of petroleum operations, global warming can be reversed without compromising healthy life styles (Chhetri and Islam, 2007a). Finally it is shown the true merit of the long-term approach that promotes sustainable techniques that are socially responsible, economically attractive, and environmentally appealing.

2. Issues in Petroleum Operations

Petroleum hydrocarbons are considered to be the backbone of the modern economy. The petroleum industry that took off from the golden era of 1930's never ceased to dominate all aspects of our society. Until now, there is no suitable alternative to fossil fuel and all trends indicate continued dominance of the petroleum industry in the foreseeable future (Service, 2005). Even though petroleum operations have been based on solid scientific excellence and engineering marvels, only recently it has been discovered that many of the practices are not environmentally sustainable. Practically all activities of hydrocarbon operations are accompanied by undesirable discharges of liquid, solid, and gaseous wastes (Khan and Islam, 2007), which have enormous impacts on the environment (Khan and Islam, 2003a; Khan and Islam, 2006b; Chhetri et al., 2008). Hence, reducing environmental impact is the most pressing issue today and many environmentalist groups are calling for curtailing petroleum operations altogether. Even though there is no appropriate tool or guideline available in achieving sustainability in this sector, there are numerous studies that criticize the petroleum sector and attempt to curtail petroleum activities (Holdway, 2002). There is clearly a need to develop a new management approach in hydrocarbon operations. The new approach should be environmentally acceptable, economically profitable and socially responsible.

The crude oil is truly a non-toxic, natural, and biodegradable product but the way it is refined is responsible for all the problems created by fossil fuel utilization. The refined oil is hard to biodegrade and is toxic to all living objects. Refining crude oil and processing natural gas use large amount of toxic chemicals and catalysts including heavy metals. These heavy metals contaminate the end products and are burnt along with the fuels producing various toxic by-products. The pathways of these toxic chemicals and catalysts show that they severely affect the environment and public health. The use of toxic catalysts creates many environmental effects that make irreversible damage to the global ecosystem. A detailed pathway analysis of formation of crude oil and the pathway of refined oil and gas clearly shows that the problem of oil and gas operation lies during synthesis or their refining.

3. Pathway Analysis of Crude and Refined Oil and Gas

3.1. Pathways of Crude Oil Formation

Crude oil is a naturally occurring liquid found in formations in the Earth consisting of a complex mixture of hydrocarbons consisting of various lengths. It contains mainly four groups of hydrocarbons among, which saturated hydrocarbon consists of straight chain of carbon atoms, aromatics consists of ring chains, asphaltenes consists of complex polycyclic hydrocarbons with complicated carbon rings and other compounds mostly are of nitrogen, sulfur and oxygen. It is believed that crude oil and natural gas are the products of huge overburden pressure and heating of organic materials over millions of years.

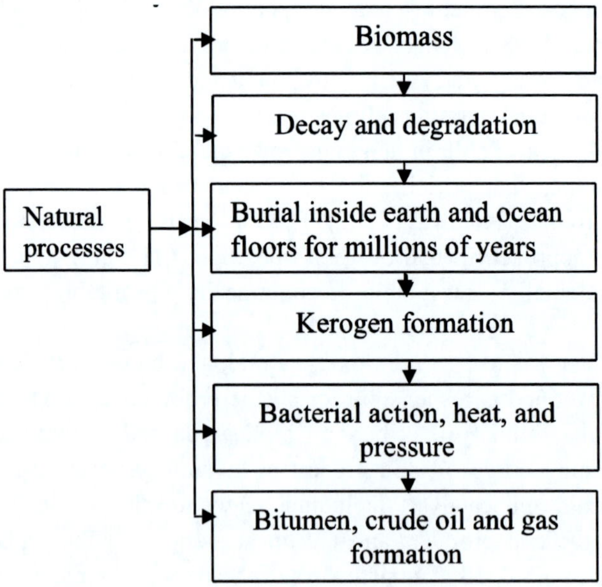

Figure 1. Crude oil formation pathway (After Chhetri and Islam, 2008a).

Crude oil and natural gases are formed as a result of the compression and heating of ancient organic materials over a long period of time. Oil, gas and coal are formed from the remains of zooplankton, algae, terrestrial plants and other organic matters after exposure to heavy pressure and temperature of Earth. These organic materials are chemically changed to kerogen. With more heat and pressure along with bacterial activities, oil and gas are formed. Figure 1 is the pathway of crude oil and gas formation. These processes are all driven by natural forces.

3.2. Pathways of Oil Refining

Fossil fuels derived from the petroleum reservoirs are refined in order to suit the various application purposes from car fuels to aeroplane and space fuels. It is a complex mixture of hydrocarbons varying in composition depending on its source. Depending on the number of carbon atoms the molecules contain and their arrangement, the hydrocarbons in the crude oil have different boiling points. In order to take the advantage of the difference in boiling point of different components in the mixture, fractional distillation is used to separate the hydrocarbons from the crude oil. Figure 2 shows general activities involved in oil refining.

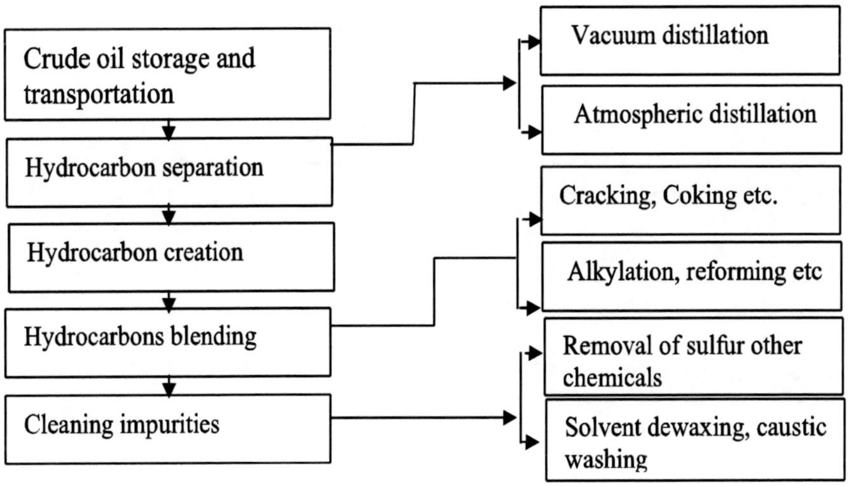

Figure 2. General activities in oil refining (Chhetri and Islam, 2008a).

Petroleum refining begins with the distillation, or fractionation of crude oils into separate hydrocarbon groups. The resultant products of petroleum are directly related to the properties of the crude processed. Most of the distillation products are further processed into more conventionally usable products changing the size and structure of the carbon chain through several processes by cracking, reforming and other conversion processes. In order to remove the impurities in the products and improve the quality, extraction, hydrotreating and sweetening are applied. Hence, an integrated refinery consists of fractionation, conversion, treatment and blending including petrochemicals processing units.

Oil refining involves the use of different types of acid catalysts along with high heat and pressure (Figure 3). The process of employing the breaking of hydrocarbon molecules is the thermal cracking. During alkylation, sulfuric acids, hydrogen fluorides, aluminum chlorides and platinum are used as catalysts. Platinum, nickel, tungsten, palladium and other catalysts are used during hydro processing. In distillation, high heat and pressure are used as catalysts. The use of these highly toxic chemicals and catalysts creates several environmental problems. Their use will contaminate the air, water and land in different ways. Use of such chemicals is not a sustainable option. The pathway analysis shows that current oil refining process is inherently unsustainable.

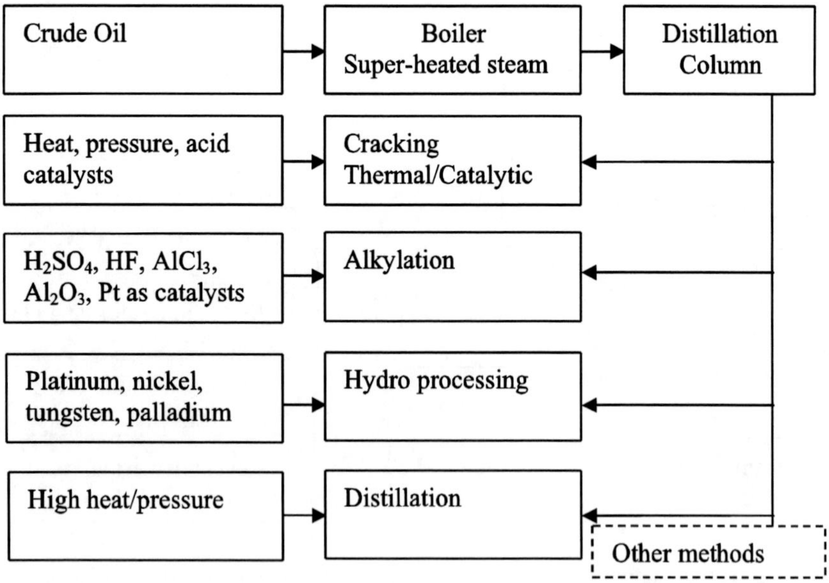

Figure 3. Pathway of oil refining process (After Chhetri et al., 2008).

3.2.1. Emission from a Refinery

Refining petroleum products emits several hazardous air toxins and particulate materials. They are produced while transferring and storage of materials and during hydrocarbon separations. Table 1 shows the emission released during the hydrocarbon separation process and handling.

Table 1. Emission from a Refinery (Environmental Defense, 2005)

Activities	Emission
Material transfer and storage	- Air release: Volatile organic compounds - Hazardous solid wastes: anthracene, benzene, 1,3-butadiene, curnene, cyclohexane, ethylbenzene, ethylene, methanol, naphthalene, phenol, PAHs, propylene, toluene, 1,2,4-trimethylbenzene, xylene
Separating hydrocarbons	- Air release: Carbon monoxide, nitrogen oxides, particulate matters, sulfur dioxide, VOCs - Hazardous solid waste: ammonia, anthracene, benzene, 1,3-butadiene, curnene, cyclohexane, ethylbenzene, ethylene, methanol, naphthalene, phenol, PAHs, propylene, toluene, 1,2,4-trimethylbenzene, xylene

Table 2. Primary wastes from oil refinery (Environmental Defense, 2005)

Cracking/coking	Alkylation and reforming	Sulfur removal
Air releases: carbon monoxide, nitrogen, oxides, particulate matter, sulfur, dioxide, VOCs	Air releases: carbon monoxide, nitrogen oxides, particulate matter, sulfur dioxide, VOCs	Air releases: carbon monoxide, nitrogen oxides, particulate, matter, sulfur dioxide, VOCs
Hazardous/solid wastes, wastewater, ammonia, anthracene, benzene, 1,3-butadiene, copper, cumene, cyclohexane, ethylbenzene, ethylene, methanol, naphthalene, nickel, phenol, PAHs, propylene, toluene, 1,2,4-trimethylbenzene, vanadium (fumes and dust), xylene	Hazardous/solid wastes: ammonia, benzene, phenol, propylene, sulfuric acid aerosols or hydrofluoric acid, toluene, xylene Wastewater	Hazardous/solid wastes: ammonia, diethanolamine, phenol, metals, Wastewater

Table 2 shows the primary waste generated from an oil refinery. In all processes, air toxics and hazardous solid materials, including volatile organic compounds are present.

3.3. Pathways of Gas Processing

Natural gas is a mixture of methane, ethane, propane, butane and other hydrocarbons, water vapor, oil and condensates, hydrogen sulfides, carbon dioxide, nitrogen, some other gases and solid particles. The free water and water vapors are corrosive to the transportation equipment. Hydrates can plug the gas accessories creating several flow problems. Other gas mixtures such as hydrogen sulfide and carbon dioxide are known to lower the heating value of natural gas by reducing its overall fuel efficiency. There are certain restrictions imposed on major transportation pipelines on the make-up of the natural gas that is allowed into the pipeline called pipe 'line quality' gas. This makes mandatory that natural gas be purified before it is sent to transportation pipelines. The gas processing is aimed at preventing corrosion, environmental and safety hazards associated with transport of natural gas.

The presence of water in natural gas creates several problems. Liquid water and natural gas can form solid ice-like hydrates that can plug valves and fittings in the pipeline (Nallinson, 2004). Natural gas containing liquid water is corrosive, especially if it contains carbon dioxide and hydrogen sulfide. Water vapor in natural gas transport systems may condense causing a sluggish flow. Hence, the removal of free water, water vapors, and condensates is a very important step during gas processing. Other impurities of natural gas, such as, carbon dioxide and hydrogen sulfide generally called as acid gases must be removed from the natural gas prior to its transportation (Chakma, 1999). Hydrogen sulfide is a toxic and corrosive gas which is rapidly oxidized to form sulfur dioxide in the atmosphere (Basu et al., 2004). Oxides of nitrogen found in traces in the natural gas may cause ozone layer depletion and global warming. Figure 5 is the schematic of general gas processing system.

Figure 4 illustrates the pathway of natural gas processing from reservoir to end uses. This figure also shows various emissions from natural gas processing from different steps. After the exploration and production, natural gas stream is sent through the processing systems.

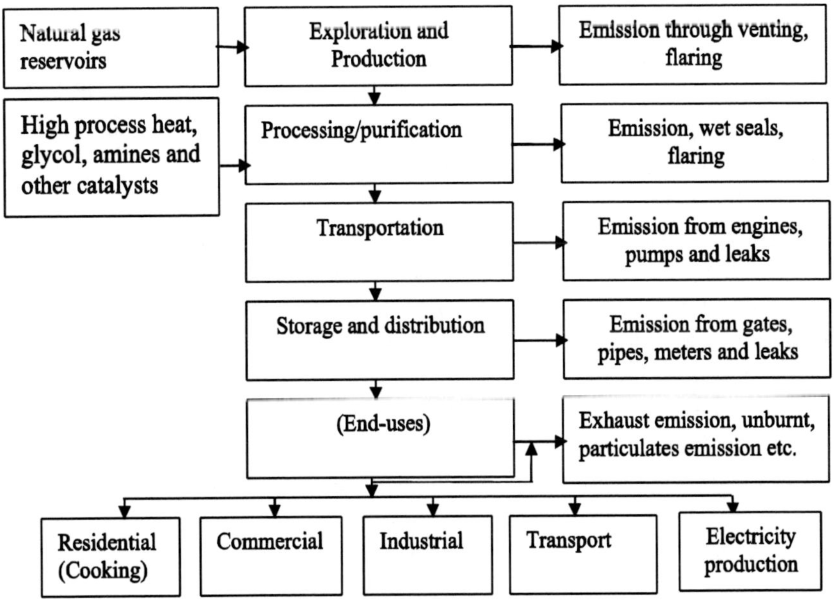

Figure 4. Natural gas "well to wheel" pathway (Redrawn from Chhetri et al., 2008).

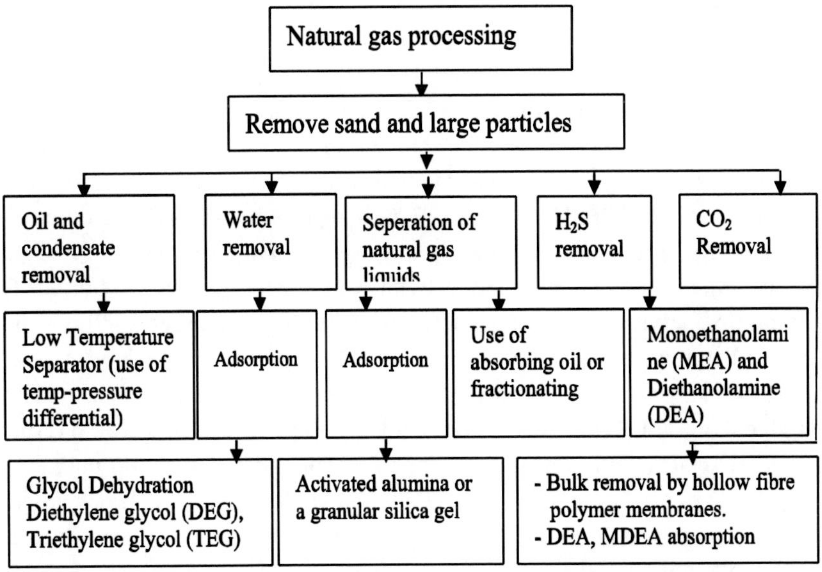

Figure 5. Natural gas processing methods (Redrawn from Chhetri and Islam, 2008).

Glycol dehydration is used for water removal from the natural gas stream. Similarly, methanolamines (MEA) and Diethanolamine (DEA) are used for removing H_2S and CO_2 from the gas streams (Figure 5). Since these chemicals are used for gas processing, it is impossible to completely free the gas from these chemicals. Glycols and amines are very toxic chemicals. Burning of ethylene glycols produces carbon monoxide (Matsuoka et al., 2005) and when the natural gas is burned in the stoves, it is possible that the emission produces carbon monoxide. Carbon monoxide is a poisonous gas and very harmful for the health and environment. Similarly, amines are also toxic chemicals and burning the gas contaminated by amines produces toxic emissions. Despite the prevalent notion that natural gas burning is clean, the emission is not free from environmental problems. It is reported that one of the highly toxic compounds released in natural gas stoves burning (LPG in stoves) is isobutane which causes hypoxia in the human body (Sugie et al, 2004).

3.3.1. Problems in Natural Gas Processing

Pathways of Glycol and Amines

Conventional natural gas processing process consists of applications of various types of chemicals and polymeric membranes. The common chemicals used to remove water, CO_2 and H_2S are Diethylene glycol (DEG) and Triethylene glycol (TEG) and Monoethanolamines (MEA), Diethanolamines (DEA) and Triethanolamine (TEA). These are synthetic chemicals and have various health and environmental impacts. Synthetic polymers used as membrane during gas processing are highly toxic and their production involves using highly toxic catalysts, chemicals, excessive heat and pressures (Chhetri et al, 2008). Hull et al. (2002) reported combustion toxicity of ethylene-vinyl acetate copolymer (EVA) reported higher yield of CO and several volatile compounds along with CO_2. Islam (2004) reported that oxidation of polymers produces more than 4000 toxic chemicals, 80 of which are known carcinogens.

Matsuoka et al. (2005) reported a study on electro oxidation of methanol and glycol and found that electro-oxidation of ethylene glycol at 400mV forms glycolate, oxalate and formate (Figure 6). The glycolate was obtained by three-electron oxidation of ethylene glycol, and was an electrochemically active product even at 400mV, which led to the further oxidation of glycolate. Oxalate was found stable, no further oxidation was seen and was termed as non poisoning path. The other product of glycol oxidation is called formate which is termed as poisoning path or CO poisoning path. The glycolate formation decreased from 40-18 % and formate increased from 15-20% between 400 and 500mV. Thus, ethylene glycol oxidation produced CO instead of CO_2 and follows the poisoning path over 500 mV. The glycol oxidation produces glycol aldehyde as intermediate products. Hence, use of these products in refining will have several impacts in the end uses, and are not sustainable at all.

Glycol ethers are known to produce toxic metabolites such as the teratogenic methoxyacetic acid during biodegradation, the biological treatment of glycol ethers can be hazardous (Fischer and Hahn, 2005). Abiotic degradation experiments with ethylene glycol showed that the by- products are monoethylether (EGME) and toxic aldehydes, e.g. methoxy acetaldehyde (MALD). Glycol passes into body by inhalation, ingestion or skin. Toxicity of ethylene glycol causes depression and kidney damage (MSDS, 2005). High concentration levels can interfere with the ability of the blood to carry oxygen causing headache and a blue color to the skin and lips (*methemoglobinemia*), collapse and even death. High exposure may affect the nervous system and may damage the red blood cells leading to anemia (low blood count). During a study of carcinogenetic and toxicity of propylene glycol on animals, the skin tumor incidence was observed (CERHR, 2003). Glycol may form toxic alcohol inside human body if ingested as fermentation may take place.

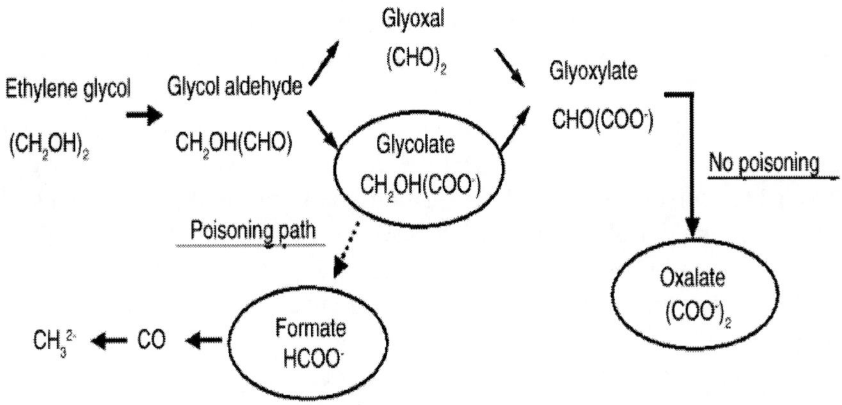

Figure 6. Ethylene Glycol Oxidation Pathway in Alkaline Solution (After Matsuoka et al., 2005).

Amines are considered to be toxic chemicals. It was reported that occupational asthma was found in people handling of a cutting fluid containing diethanolamine (Piipari et al., 1998). Toninello (2006) reported that the oxidation products of some biogenic amines appear to be also carcinogenic. DEA also reversibly inhibits phosphatidylcholine synthesis by blocking choline uptake (Lehman-McKeeman and Gamsky, 1999). Systemic toxicity occurs in many tissue types including the nervous system, liver, kidney, and blood system that may cause increased blood pressure, diuresis, salivation, and pupillary dilation. Diethanolamine causes mild skin irritation to the rabbit at concentrations above 5%, and severe ocular irritation at concentrations above 50% (Beyer *et al.,* 1983). Ingestion of diethylamine causes severe gastrointestinal pain, vomiting, and diarrhea, and may result in perforation of the stomach possibly due to the oxidation products and fermentation products.

4. Critical Evaluation of Current Petroleum Practices

In very short historical time (relative to the history of the environment) the oil and gas industry has become one of the world's largest economic sectors, a powerful globalizing force with far reaching impacts on the entire planet humans share with the rest of the natural world. Decades of the continuous growth of oil and gas operations have changed, in some places transformed the natural environment and the way humans have traditionally organized themselves. The petroleum sectors draw huge public attention due to their environmental consequences. All stages of oil and gas operations generate a variety of solids, liquids and gaseous wastes (Currie and Isaacs, 2005; Wenger et al., 2004; Khan and Islam, 2003b; Veil, 2002; de Groot, 1996; Holdway, 2002) harmful to human and the natural environment. Figure 7 shows the current technological practices are focused on short-term, linearized solutions that are also aphenomenal. As a result, technological disaster prevails practically in every aspect of the post-renaissance era. Petroleum practices are considered to be the driver of today's society. Here, the modern development is essentially dependent on artificial products and processes. The *modus operandi* was summarized by Zatzman (2007). He called the post renaissance transition the honey-sugar-saccharine-aspartame (HSSA) syndrome.

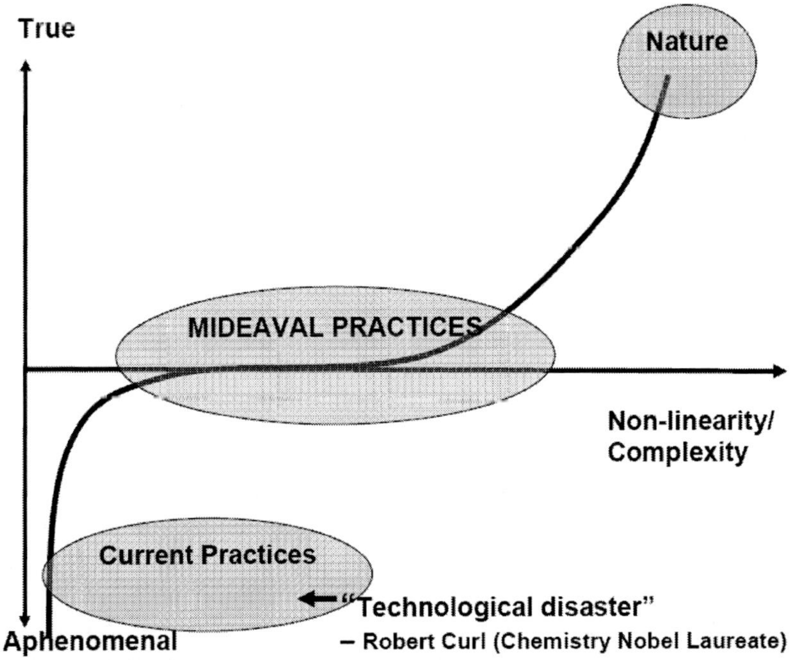

Figure 7. Schematic showing the position of current technological practices related to natural practices.

In this allegorical transition, honey with real source and real process) has been systematically replaced by Aspartame that has both source and pathway that are highly artificial. Such transition, he argued was allowed to take place because the profit margin increases with processing. This sets in motion the technology development mode that Nobel Laureate in Chemistry-Robert Curl called "technological disaster".

By now, it has been recognized the unpleasant truth that the present natural resource management regime governing activities such as petroleum operations has failed to ensure environmental safety and ecosystem integrity. The main reason for this failure is that the existing management scheme is not sustainable (Khan and Islam 2007a). Under the present management approach, development activities are allowed so long as they promise economic benefit. Once that is likely, management guidelines are set to justify the project's acceptance.

Sustainable petroleum operations development requires a sustainable supply of clean and affordable energy resources that do not cause negative environmental, economic, and social consequences (Dincer and Rosen 2004,

2005). In addition, it should consider a holistic approach where the whole system will be considered instead of just one sector at a time (Mehedi et al. 2007a, 2007b). Recently, Khan and Islam (2005b; 2006a) and Khan et al. (2005) developed an innovative criterion for achieving true sustainability in technological development. This criterion can be applied effectively to offshore technological development. New technology should have the potential to be efficient and functional far into the future in order to ensure true sustainability. Sustainable development is seen as having four elements–economic, social, environmental, and technological.

4.1. Management

Conventional management of petroleum operations are being challenged due to the environmental damages caused by its operations. Moreover, the technical picture of the petroleum operations and management is very grim (Deakin and Konzelmann, 2004). The ecological impacts of petroleum discharges including habitat destruction and fragmentation, recognized as major concerns associated with petroleum and natural gas developments in both terrestrial and aquatic environments. There is clearly a need to develop a new management approach in hydrocarbon operations. This approach will have to be environmentally acceptable, economically profitable and socially responsible. These problems might be solved/overcome by the application of new technologies which guarantee sustainability.

Figure 8 shows the different phases of petroleum operations which are seismic, drilling, production, transportation and processing and decommissioning, and their associated wastes generation and energy consumption. Various types of waste from ships, emission of CO_2, human related waste, drilling mud, produced water, radioactive materials, oil spills, release of injected chemicals, toxic release used as corrosion inhibitors, metals and scraps, flare etc are produced during the petroleum operations. Even though, petroleum companies make billions of dollars profit in their operations each year, these companies take no responsibilities of various waste generated. Hence, in overall, the society is deteriorated due to such environmental damages. Until the mesh created by petroleum operations are rendered to environmental friendly operation, the society as a whole will not be benefited from such valuable natural resources.

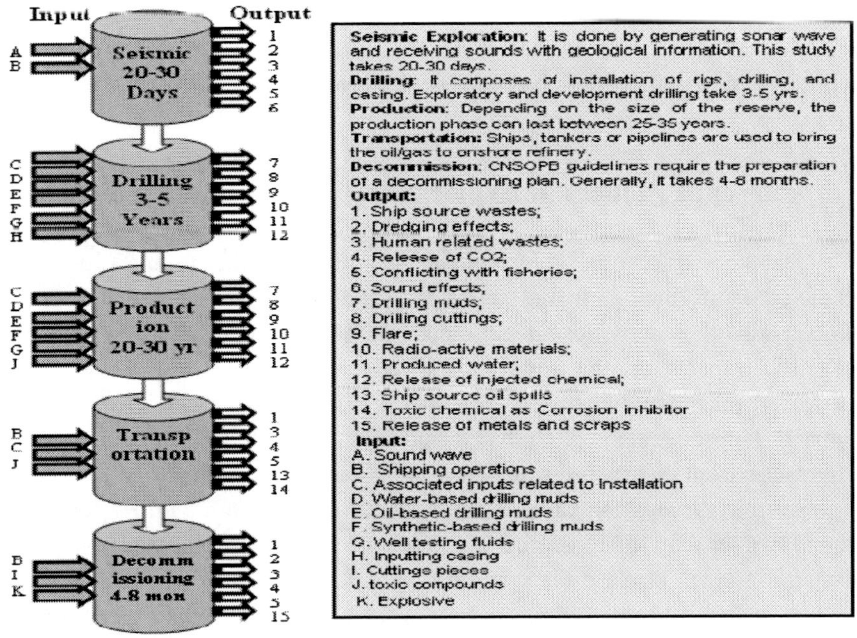

Figure 8. Different phases of petroleum operations which are seismic, drilling, production, transportation and processing and decommissioning, and their associated wastes generation and energy consumption (Khan and Islam, 2006a).

Recently, Khan et al. (2005) and Khan and Islam (2005) introduced a new approach by means of which it is possible to develop a truly sustainable technology. Under this approach, the temporal factor is considered the prime indicator in sustainable technology development. Khan and Islam (2006a; 2007) discussed some implications of how the current management model for exploring, drilling, managing wastes, refining and transporting, and using the by-products of petroleum has been lacking in foresight, and suggests the beginnings of a new management approach.

A common practice among all oil producing companies is to burn off any unwanted gas that liberates from oil during production. This process ensures the safety of the rig by reducing the pressures in the system that result from gas liberation. This gas is of low quality and contains many impurities and by burning off the unwanted gas, toxic particles are released into the atmosphere. Acid rain, caused by sulfur oxides in the atmosphere, is one of the main environmental hazards resulting from this process. Moreover, flaring of natural gas accounts for approximately a quarter of the petroleum industries emissions (UKOOA, 2003). At present, flaring of gases onsite and disposing of liquid and solid containing less than a certain concentration of hydrocarbon are allowed. It is expected, however, that more stringent operating conditions will be needed to meet the objectives set by Kyoto protocols.

4.2. Current Practices in Exploration, Drilling and Production

Seismic exploration is examined for the preliminary investigation of geological information in the study area and is considered to be the safest among all other activities in petroleum operations, with little or negligible negative impacts on the environment (Diviacco 2005; Davis et al. 1998). However, recent studies have shown that it has several adverse environmental impacts (Jepson et al. 2003; Khan and Islam 2007). Most of the negative effects are from the intense sound generated during the survey.

Seismic surveys can cause direct physical damage to a fish. High pressure sound waves can damage the hearing system, swim bladders, and other tissues/systems. These effects might not directly kill the fish, but they may lead to reduced fitness, which increases their susceptibility to predation and decreases their ability to carry out important life processes. There might be indirect effects from seismic operations. If the seismic operation disturbs the food chain/web, then it will cause adverse impacts on fish and total fisheries. The physical and behavioral effects on fish from seismic operations are discussed in the following sections. It has also been reported that seismic surveys cause behavioral effects among fish. For example, startle response, change in swimming patterns (potentially including change in swimming speed and directional orientation), and change in vertical distribution are some of the effects. These effects are expected to be short-term, with duration of effect less than or equal to the duration of exposure, they are expected to vary between species and individuals, and be dependent on properties of received sound. The ecological significance of such effects is expected to be low, except where they influence reproductive activity.

Some studies of the effects of seismic sound on eggs and larvae or on zooplankton were found. Other studies showed that exposure to sound may arrest development of eggs, and cause developmental anomalies in a small proportion of exposed eggs and/or larvae; however these results occurred at numbers of exposures much higher than are likely to occur during field operation conditions, and at sound intensities that only occur within a few meters of the sound source. In general, the magnitude of mortality of eggs or larvae that could result from exposure to seismic sound predicted by models would be far below that which would be expected to affect populations. Similar physical, behavioral and physiological effects in the invertebrates are also reported. Marine turtles and mammals are also significantly affected due to seismic activities.

Bjorndalen et al. (2005) developed a novel approach to avoid flaring during petroleum operations. Petroleum products contain materials in various phases. Solids in the form of fines, liquid hydrocarbon, carbon dioxide, and hydrogen sulfide are among the many substances found in the products. According to Bjorndalen et al. (2005), by separating these components through the following steps, no-flare oil production can be established (Figure 9). Simply by avoiding flaring, over 30% of pollution created by petroleum operation can be reduced. Once the components for no-flaring have been fulfilled, value added end products can be developed. For example, the solids can be used for minerals, the brine can be purified, and the low-quality gas can be re-injected into the reservoir for EOR.

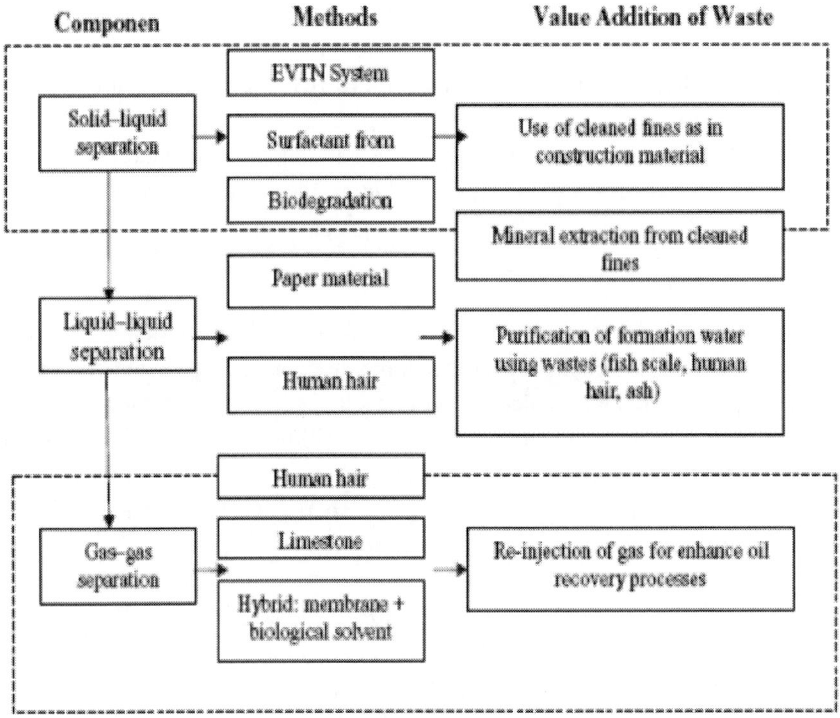

Figure 9. Breakdown of no-flaring method (Bjorndalen et al. 2005).

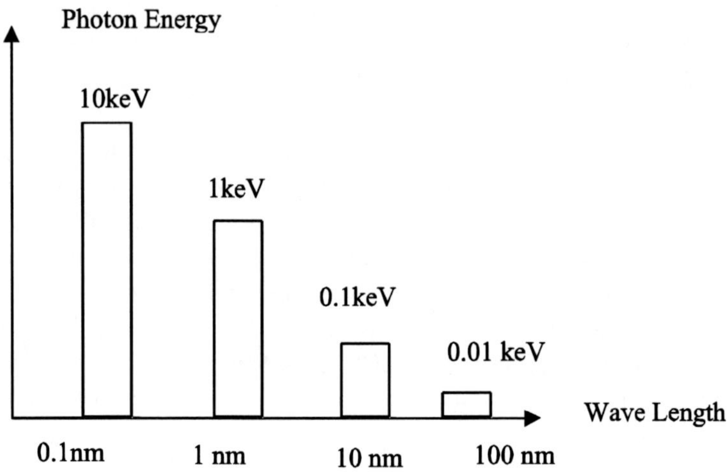

Figure 10. Schematic of wave length and energy level of photon.

The use of drilling technologies for exploration has been of great concerns recently. Himpsel (2007) presented a correlation between the energy levels and the wave length of photon energy (Figure 10). It is shown that the energy level of photon decreases with the increase in wave length. Hence, it is obvious that more energy needs to be spent to generate energy of higher intensity such as drilling in the petroleum operations. In addition, more environmental hazard is created due to the production of high intensity energy sources such as laser drillings.

Table 3 shows the quantum energy level of various radiation sources. The γ-rays which have the least wave length have the highest quantum energy levels. In terms of intensity, γ-rays have highest energy intensity among others. More energy is needed to produce this radiation whether to use for drilling or any other application. Moreover, drilling and production activities have also adverse effects on the environment in several ways. For example, blow-out and flaring of produced gas waste energy, carbon dioxide emissions into the atmosphere, and careless disposal of drilling mud and other oily materials, can have a toxic effect on terrestrial and marine life. Before drilling and production operations are allowed to go ahead, the Valued Ecosystem Component (VEC) level impact assessment should be done to establish the ecological and environmental conditions of the area proposed for development and assess the risks to the environment from the development.

Table 3. Wave length and quantum energy levels of different radiation sources (Website 3)

Radiation	Wave length	Quantum energy
Infrared	1 mm - 750 nm	0.0012 - 1.65 eV
Visible	750 - 400 nm	1.65 - 3.1 eV
Ultraviolct	400 nm - 10 nm	3.1 - 124 eV
X-rays	10 nm	124 eV
γ-rays	10^{-12} m	1 MeV

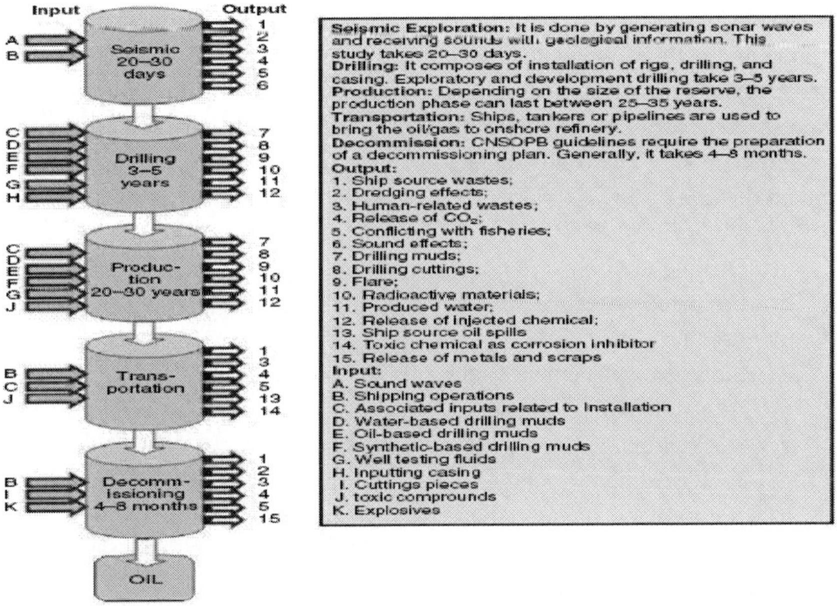

Figure 11. Supply chain of petroleum operations (Khan and Islam, 2006a).

4.3. Challenges in Waste Management

Drilling and production phases are the most waste generating phase in petroleum operations. Drilling mud are condensed liquids that may be oil- or synthetic-based wastes, and contain a variety of chemical additives and heavy minerals that are circulated through the drilling pipe to perform a number of functions. These functions include cleaning and conditioning the hole, maintaining hydrostatic pressure in the well, lubrication of the drill bit and counterbalance formation pressure, removal of the drill cuttings, and stabilization the wall of the drilling hole. Water-based muds (WBMs) are a complex blend of water and bentonite. Oil-based muds (OBMs) are composed of mineral oils, barite, mineral oil, and chemical additives. Typically a single well may lead to 1000–6000 m3 of cuttings and muds depending on the nature of cuttings, well depths, and rock types (CEF, 1998). A production platform generally consists of 12 wells, which may generate (62 × 5000 m3) 60,000 m3 of wastes (Patin 1999; CEF 1998). Figure 11 shows the supply chain of petroleum operation indicating the type of wastes generated.

The current challenge of petroleum operation is how to minimize the petroleum wastes and its impact in the long-term. Conventional drilling and production methods generate an enormous amount of wastes (Veil 1998; EPA, 2000). Existing management practices are mainly focused to achieve sectoral success and are not coordinated with other operations surrounding the development site. The following are the major wastes generated during drilling and production.

a) Drilling muds
b) Produced water
c) Produced sand
d) Storage displacement water
e) Bilge and ballast water
f) Deck drainage
g) Well treatment fluids
h) Naturally occurring radioactive materials
i) Cooling water
j) Desalination brine
k) Other assorted wastes

4.4. Problems in Transportation Operations

The most significant problems in production and transportation operations are reported by Khan and Islam (2006a). Toxic means are used to control corrosion. Billions of dollars are spent to add toxic agents so that microbial corrosion can be prevented. Other applications include the use of toxic paints, cathodic protection, etc. all causing irreversible damage to the environment. In addition to this, huge amount of toxic chemicals (at times up to 40% of the gas stream) are injected in order to reduce moisture content of a gas stream in order to prevent hydrate formation. Even if 1% of these chemical remain in the gas stream, the gas becomes vulnerable to severe toxicity, particularly when it is burnt, contaminating the entire pathway of gas usage (Chhetri and Islam, 2008). Toxic solvents are used for preventing asphaltene plugging problems. Toxic resins are used for sand consolidation to preventing sand production.

5. Greening of Petroleum Operations

5.1. Effective Separation of Solid from Liquid

Organic waste products such as, cattle manure, slaughter house waste, vegetable waste, fruit peels and pits, dried leaves and natural fibers, wood ash, natural rocks (limestone, zeolite, siltstone, etc.) are all viable options for the separation of fines and oil. Recently, a patent has been issued to Titanium Corporation for separation of oil from sand tailings. However, this technique uses chemical systems, such as NaOH and H_2O_2, which are both expensive and environmentally hostile. Drilling wastes have been found to be beneficial in highway construction (Wasiuddin et al., 2002b). Studies have shown that the tailings from oil sands are high in various mineral contents. To extract the minerals, usually the chemical treatment is used to modify the surface of minerals. Treatment with a solution derived from natural material has a great potential (Islam, 2003). As well, microwave heating has the potential to enhance selective floatability of different particles. This aspect has been studied by Gunal and Islam (2000). Temperature can be a major factor in the reaction kinetics of a biological solvent with mineral surfaces. Various metals respond in a different manner under microwave condition, which can make significant change in floatability. The recovery process can be completed through transferring the microwave-treated fines to a flotation chamber (Henda et al., 2005). Finally, atomic absorption spectrometer can characterize the flotation products.

Application of bio membranes to separate solid from liquid has also given considerable attention recently. Even though, synthetic membranes are being used for some application, they are highly energy intensive to produce, toxic and costly.

5.2. Effective Separation of Liquid from Liquid

The current practice involves separation of oil and water and the water is disposed as long as it has a hydrocarbon concentration below an allowable limit. However, it is expected that such practice cannot be sustained and further purification of water is necessary. Consequently, this task involves on both separation of oil and water and heavy metals removal from the water. The oil-water emulsion has been produced and it was found that the selected paper fibre material gives 98-99% recovery of oil without producing any water (Khan and Islam, 2006c). An emulsion made up of varying ratios of oil and water was utilized in different sets of experiments. Almost all the oil-water ratios gave the same separation efficiency with the material used. The mentioned paper material, which is made of long fibrous wood pulp treated with water proofing material as filtering medium. Water proofing agent for paper used in these experiments is "Rosin Soap" (rosin solution treated with caustic soda). This soap is then treated with alum to keep the ph of the solution within a range of 4~5. Cellulose present in the paper reacts reversibly with rosin soap in presence of alum and form a chemical coating around the fibrous structure and acts as coating to prevent water to seep through it. This coating allows long chain oil molecules to pass making the paper a good conductor for oil stream. Because of the reversible reaction, it was also observed the performance of filter medium increases with the increased acidity of the emulsion and vice versa. As well, the filter medium is durable to make its continuous use for a long time keeping the cost of replacement and production cut-down. The material used as filtering medium is environmental friendly and useful for down-hole conditions. Further, other paper-equivalent alternate materials, which are inexpensive and down-hole environment appealing, can be used for oil-water separation. Human hair has been proven to be effective in separation of oil and water as well as heavy metal removal.

Similarly, natural zeolites can also effectively function to separate liquid-liquid from different solutions. Such zeolites can adsorb some liquid leaving others to separate depending on their molecular weight.

5.3. Effective Separation of Gas from Gas

Most of the hydrocarbons found in natural gas wells are complex mixtures of hundreds of different compounds. A typical natural gas stream is a mixture of methane, ethane, propane, butane and other hydrocarbons, water vapor, oil and condensates, hydrogen sulfides, carbon dioxide, nitrogen, some other gases and solid particles. The free water and water vapors are corrosive to the transportation equipment. Hydrates can plug the gas accessories creating several flow problems. Other gas mixtures such as hydrogen sulfide and carbon dioxide are known to lower the heating value of natural gas by reducing its overall fuel efficiency. There are certain restrictions imposed on major transportation pipelines on the make-up of the natural gas that is allowed into the pipeline called pipe 'line quality' gas. This makes mandatory that natural gas is purified before it is sent to transportation pipelines. The gas processing is aimed at preventing corrosion, environmental and safety hazards associated with transport of natural gas.

The presence of water in natural gas creates several problems. Liquid water and natural gas can form solid ice-like hydrates that can plug valves and fittings in the pipeline (Nallinson, 2004). Natural gas containing liquid water is corrosive, especially if it contains carbon dioxide and hydrogen sulfide. Water vapor in natural gas transport systems may condense causing a sluggish flow. Hence, the removal of free water, water vapors, and condensates is a very important step during gas processing. Other impurities of natural gas, such as, carbon dioxide and hydrogen sulfide generally called as acid gases must be removed from the natural gas prior to its transportation (Chakma, 1999). Hydrogen sulfide is a toxic and corrosive gas which is rapidly oxidized to form sulfur dioxide in the atmosphere (Basu et al., 2004). Oxides of nitrogen found in traces in the natural gas may cause ozone layer depletion and global warming. Hence, an environment-friendly gas processing is essential in order for greening the petroleum operations

5.3.1. Natural Substitutes for Gas Processing Chemicals (Glycol and Amines)

Glycol is one of the most important chemical used during the dehydration of natural gas. In search of the cheapest, most abundantly available and cheap material, clay has been considered as one of the bets substitute of toxic glycol. Clay is a porous material containing various minerals such as silica, alumina, and several others. Low et al. (2003) reported that the water absorption characteristics of sintered sawdust clay can be modified by the addition of saw dust particles to the clay. The dry clay as a plaster has water absorption coefficient of 0.067-0.075 ($kg/m^2S^{1/2}$) where weight of water absorbed is in kg, surface area in square meter and time in second (Website 1). The preliminary experimental result has indicated that clay can absorb considerable amount of water vapor and can be efficiently used in dehydration of natural gas (Figure 12). Moreover, glycol can be obtained from some natural source, which is not toxic as synthetic glycol. Glycol can be extracted from Tricholoma Matsutake (mushroom) which is an edible fungus (Ahn and Lee, 1986). Ethylene glycol is also found as a metabolite of ethylene which regulates the natural growth of the plant (Blomstrom and Beyer, 1980). Orange peel oils can replace this synthetic glycol. These natural glycols derived without using non organic chemicals can replace the synthetic glycols. Recent work of Miralai et al.(2006) have demonstrated that such considerations are vital.

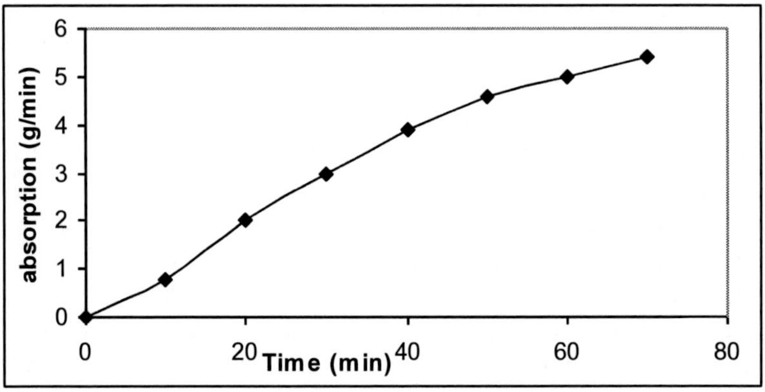

Figure 12. Water vapor absorption by Nova Scotia clay (Chhetri and Islam, 2008).

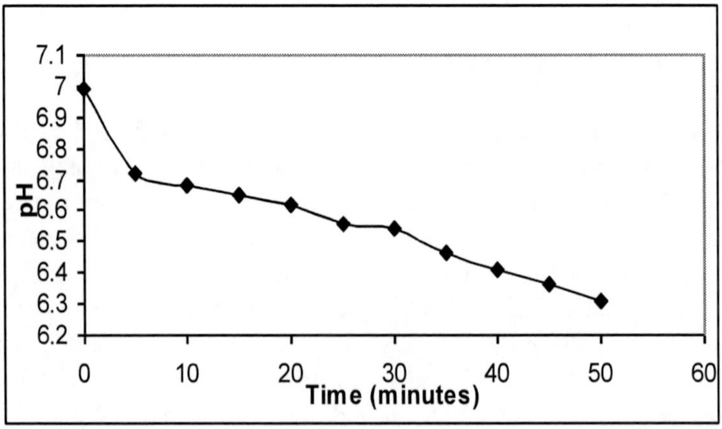

Figure 13. Decrease of pH with time due to sulfur absorption in de-ionized water (Chhetri and Islam, 2008).

Amines are used in natural gas processing to remove H_2S and CO_2. Monoethanolamine (MEA), DEA and TEA are the members of alkanolamine compound. These are synthetic chemicals the toxicity of which has been discussed earlier. If these chemicals are extracted from natural sources, such toxicity is not expected. Monoethanolamine is found in the hemp oil which is extracted from the seeds of hemp (Cannabis Sativa) plant. 100 grams of hemp oil contain 0.55 mg of Monoethanolamine (Website 2). Moreover, an experimental study showed that olive oil and waste vegetable oil can absorb sulfur dioxide. Figure 13 indicates the decrease in pH of de-ionized water with time. This could be a good model to remove sulfur compounds from the natural gas streams. Calcium hydroxides can also be utilized to remove CO_2 from the natural gas.

5.3.2. Membranes and Absorbents

Various types of synthetic membranes are in use for the gas separation. Some of them are liquid membranes and some are polymeric. Liquid membranes operate by immobilizing a liquid solvent in a microporous filter or between polymer layers. A high degree of solute removal can be obtained when using chemical solvents. When the gas or solute reacts with the liquid solvent in the membrane, the result is an increased liquid phase diffusivity. This leads to an increase in the overall flux of the solute. Furthermore, solvents can be chosen to selectively remove a single solute from a gas stream to improve selectivity (Astrita et al. 1983). Saha and Chakma (1992) suggested the attachment of a liquid membrane in a microporous polymeric membrane. They immobilized mixtures of various amines such as monoethanolamine (MEA), diethanolamine (DEA), amino-methyl-propanol (AMP), and polyethylene glycol (PEG) in a microporous polypropylene film and placed it in a permeator. They tested the mechanism for the separation of carbon dioxide from some hydrocarbon gases and obtained separation factors as high as 145.

Polymeric membranes have been developed for a variety of industrial applications, including gas separation. For gas separation, the selectivity and permeability of the membrane material determines the efficiency of the gas separation process. Based on flux density and selectivity, a membrane can be classified broadly into two classes, porous and nonporous. A porous membrane is a rigid, highly voided structure with randomly distributed interconnected pores. The separation of materials by a porous membrane is mainly a function of the permeate character and membrane properties, such as the molecular size of the membrane polymer, pore size, and pore-size distribution. A porous membrane is similar in its structure and function to the conventional filter. In general, only those molecules that differ considerably in size can be separated effectively by microporous membranes. Porous membranes for gas separation do exhibit high levels of flux but inherit low selectivity values. However, synthetic membranes are not as environment-friendly as biodegradable bio membranes.

The efficiency of polymeric membranes decreases with time due to fouling, compaction, chemical degradation, and thermal instability. Because of this limited thermal stability and susceptibility to abrasion and chemical attack, polymeric membranes have found application in separation processes where hot reactive gases are encountered. This has resulted in a shift of interest toward inorganic membranes.

Inorganic membranes are increasingly being explored to separate gas mixtures. Besides having appreciable thermal and chemical stability, inorganic membranes have much higher gas fluxes when compared to polymeric membranes. There are basically two types of inorganic membranes, dense (nonporous) and porous. Examples of commercial porous inorganic membranes are ceramic membranes, such as alumina, silica, titanium, glass, and porous metals, such as stainless steel and silver. These membranes are characterized by high permeabilities and low selectivities. Dense inorganic membranes are specific in their separation behaviours, for example, Pd-metal based membranes are hydrogen specific and metal oxide membranes are oxygen specific. Palladium and its alloys have been studied extensively as potential membrane materials. Air Products and Chemical Inc. developed the Selective Surface Flow (SSF) membrane. It consists of a thin layer (2–3 m) of nano-porous carbon supported on a macroporous alumina tube (Rao et al. 1992). The effective pore diameter of the carbon matrix is 5–7 A (Rao and Sircar 1996). The membrane separates the components of a gas mixture by a selective adsorption-surface diffusion-desorption mechanism (Rao and Sircar 1993).

A variety of bio-membranes are also in use today. These membranes such as human hair can be used instead of synthetic membranes for gas-gas separation (Basu et al., 2004; Akhter, 2002). The use of human hair as biomembrane has been illustrated by (Khan and Islam, 2006a). Initial results indicated that human hairs have characteristics similar to hollow fibre cylinders, but are even more effective because of the flexible nature and a texture that can allow the use of a hybrid system through solvent absortion along with mechanical separation. Natural absorbents such as silica gels can also be used for absorbing various contaminants from the natural gas stream.

Khan and Islam (2006a) showed that synthetic membranes can be replaced by simple paper membranes for oil water separation. Moreover, limestone has the potential to separate sulfur dioxide from natural gas (Akhter, 2002). When caustic soda is combined with wood ash, it was found to be an alternative to zeolite. Since caustic soda is a chemical, waste materials such as okra extract can be a good substitute. The same technique can be used with any type of exhaust, large (power plants) or small (cars). Once the gas is separated, low quality gas can then be injected into the reservoir for enhanced oil recovery technique. This will enhance the system efficiency. Moreover, low quality can be converted into power by a turbine. Bjorndalen et al.(2005) developed a comprehensive scheme for the separation of petroleum products in different form using novel materials with value addition of the by-products.

5.4. A Novel Desalination Technique

Management of produced water during petroleum operations offers a unique challenge. The concentration of this water is very high and cannot be disposed of outside. In order to bring down the concentration, expensive and energy-intensive techniques are being practiced. Recently, Khan et al. (2006b; 2006c) have developed a novel desalination technique that can be characterized as totally environment-friendly process. This process uses no non-organic chemical (e.g. membrane, additives). This process relies on the following chemical reactions in four stages:

(1) saline water + CO_2 + NH_3 → (2) precipitates (valuable chemicals) + desalinated water → (3) plant growth in solar aquarium → (4) further desalination

This process is a significant improvement over an existing US patent. The improvements are in the following areas:

- CO_2 source is exhaust of a power plant (negative cost)
- NH_3 source is sewage water (negative cost + the advantage of organic origin)
- Addition of plant growth in solar aquarium (emulating the world's first and the biggest solar aquarium in New Brunswick, Canada).

This process works very well for general desalination involving sea water. However, for produced water from petroleum formations, it is common to encounter salt concentration much higher than sea water. For this, water plant growth (Stage 3 above) is not possible because the salt concentration is too high for plant growth. In addition, even Stage 1 doesn't function properly because chemical reactions slow down at high salt concentrations. This process can be enhanced by adding an additional stage. The new process should function as:

(1) Saline water + ethyl alcohol → (2) saline water + CO_2 + NH_3 → (3) precipitates (valuable chemicals) + desalinated water → (4) plant growth in solar aquarium → (5) further desalination

Care must be taken, however, to avoid using non-organic ethyl alcohol. Further value addition can be performed if the ethyl alcohol is extracted from fermented waste organic materials.

5.5. A Novel Refining Technique

Khan and Islam (2007) have identified the following sources of toxicity in conventional petroleum refining:

- Use of toxic catalyst
- Use of artificial heat (e.g. combustion, electrical, nuclear)

The use of toxic catalysts contaminates the pathway irreversibly. These catalysts should be replaced by natural performance enhancers. Such practices have been proposed by Chhetri and Islam (2008c) in the context of biodiesel. In this proposed project, research will be performed in order to introduce catalysts that are available in their natural state. This will make the process environmentally acceptable and will reduce to the cost very significantly.

The problem associated with efficiency is often covered up by citing local efficiency of a single component (Islam et al., 2006). When global efficiency is considered, artificial heating proves to be utterly inefficient (Khan et al., 2006b; Chhetri and Islam, 2008). Recently, Khan and Islam (2006b) have demonstrated that direct heating with solar energy (enhanced by a parabolic collector) can be very effective and environmentally sustainable. They achieved up to 75% of global efficiency as compared to some 15% efficiency when solar energy is used through electricity conversion. They also discovered that the temperature generated by the solar collector can be quite high, even for cold countries. In hot climates, the temperature can exceed 300° C, making it suitable for thermal cracking of crude oil. In this project, the design of a direct heating refinery with natural catalysts will be completed. Note that the direct solar heating or wind energy doesn't involve the conversion into electricity that would otherwise introduce toxic battery cells and would also make the overall process very low in efficiency.

5.5.1. Use of Solid Acid Catalyst for Alkylation

Refiners typically use either hydrofluoric acid (HF), which can be deadly if spilled, or sulfuric acid, which is also toxic and increasingly costly to recycle. Refineries can use a solid acid catalyst, unsupported and supported forms of heteropolyacids and their cation exchanged salts, which has recently proved effective in refinery alkylation. A solid acid catalyst for alkylation is less widely dispersed into the environment compared to HF. Changing to a solid acid catalyst for alkylation would also promote more safety at a refinery. Solid acid catalysts are an environment-friendly replacement for liquid acids, used in many significant reactions, including alkylation of light hydrocarbon gases to form iso-octane (alkylate) used in reformulated gasoline. Use of organic acids and enzymes for various reactions is to be promoted.

5.5.2. Use of Nature Based Less or Non Toxic Catalyst

The catalysts that are in use today are very toxic and wasted after a series of use. This will create pollution to the environment, so using catalysts with fewer toxic materials significantly reduces pollution. The use of Nature-based catalysts such as zeolites, alumina, and silica should be promoted. Various biocatalyst and enzymes, which are nontoxic and from renewable origin, are to be considered for future use.

5.5.3. Use of Bacteria to Breakdown Heavier Hydrocarbons to Lighter Ones

Since the formation of crude oil is the decomposition of biomass by bacteria at high temperature and pressure, there must be some bacteria that can effectively break down the crude oil into lighter products. A series of investigations are necessary to observe the effect of bacteria on the crude oil.

5.5.4. Zero Waste Approach

Any chemical or industrial process should be designed in such a way that a closeloop system is maintained so that all wastes are absorbed within the assimilative capacity of the Earth. Recycling the resources and using them for alternative use will lead to a zero waste approach (Bjorndalen et al., 2005).

5.5.5. Use of Cleaner Crude Oil

Crude oil itself is comparatively cleaner than distillates as it contains less sulfur and toxic metals. The use of crude oil for various applications is to be promoted. This will not only help to maintain the environment because of its less toxic nature but also be less costly as it avoids expensive catalytic refining processes. Recently, the direct use of crude oil is of great interest.

Several studies have been conducted to investigate the electricity generation from saw dust (Sweis, 2004; Venkataraman et al., 2004; Calle et al., 2005). Figure 14 shows the schematic of a scaled model developed by Prof. Islam in collaboration with Veridity Environmental Technologies (Halifax, Nova Scotia). A raw sawdust silo is equipped with a powered auger sawdust feeder. The saw dust is inserted inside another feeding chamber that is equipped with a powered grinder that pulverizes sawdust into wood flour. The chamber is attached to a heat exchanger that dries the saw dust before it enters into the grinder. The wood flour is fed into the combustion chamber with a powered auger wood flour feeder. The pulverization of sawdust increases the surface area of the particles very significantly. The additional energy required to run the feeder and the grinder is provided by the electricity generated by the generator itself, requiring no additional energy investment. In addition the pulverization chamber is also used to dry the saw dust. The removal of moisture increases flammability of the feedstock. The combustion chamber itself is equipped with a start-up fuel injector that uses biofuel. Note that initial temperature required to startup the combustion chamber is quite high and cannot be achieved without a liquid fuel. The exhaust of the combustion chamber is circulated through a heat exchanger in order to dry sawdust prior to pulverization. As the combustion fluids escape the combustion chamber, they turn the drive shaft blades rotate to turn the drive shaft, which in turn, turns the compressor turbine blades. The power generator is placed directly under the main drive shaft.

Fernandes and Brooks (2003) compared black carbon (BC) derived from various sources. One interesting feature of this study was that they studied the impact of different sources on the composition, extractability and bioavailability of resulting BC. By using molecular fingerprints, the concluded that fossil BC may be more refractory than plant derived BC. This is an important finding as only recently there has been some advocacy that BC from fossil fuel may have cooling effect, nullifying the contention that fossil fuel burning is the biggest contributor to global warming.

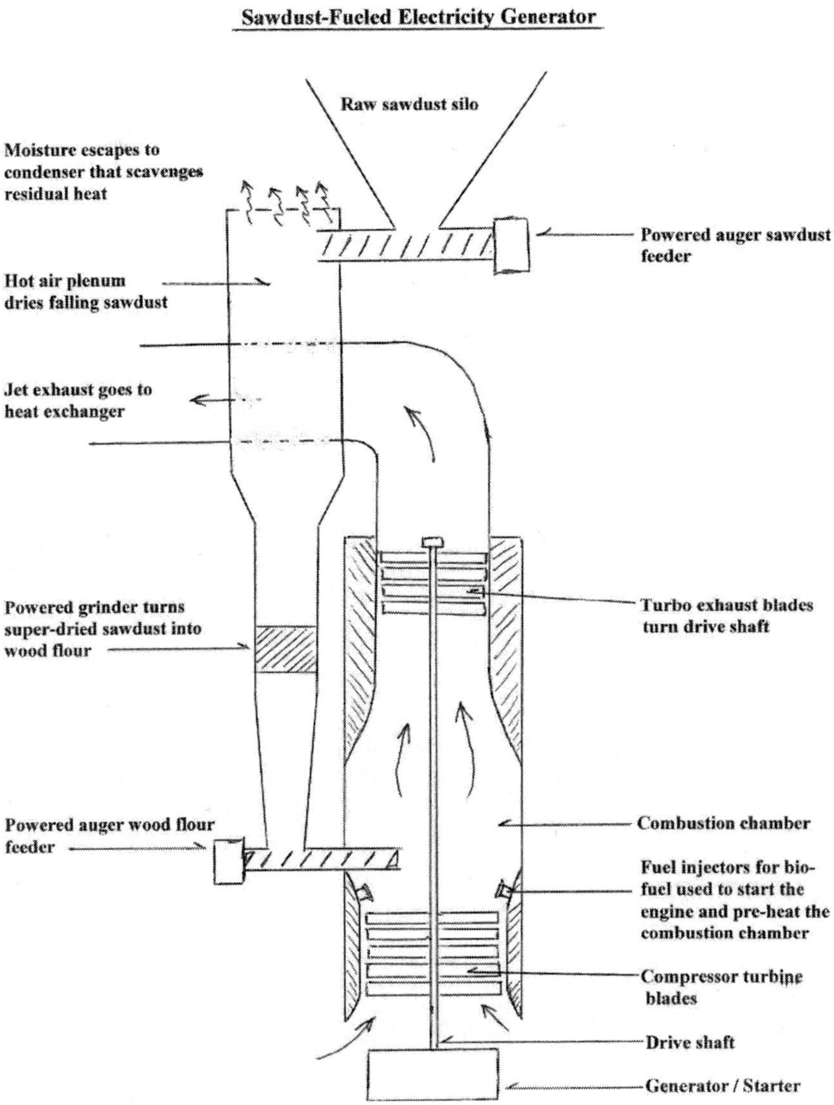

Figure 14. Schematic of sawdust fuelled electricity generator.

It is possible that BC from fossil fuel has higher refractory ability, however, there is no study available to date to quantify the cooling effect and to determine the overall effect of BC from fossil fuel. As for the other effects, BC from fossil appear to be on the harmful side as compared to BC from organic matters. For instance, vegetarian fire residues, straw ash and wood charcoals had only residual concentrations of n-alkanes (<9 μg/g) and polyclyclic aromatic (PAHs) of less than 0.2 μg/g. These concentrations compared with Diesel soot, urban dust and chimney soot PAH concentrations of greater than 8 μg/g and n-alkanes greater than 20 μg/g.

This design shows that even the solid fuels can be used to produce electricity at high efficiencies. Burning of saw dust produces fresh carbon dioxide compared burning of fossil fuels which produces older carbon dioxide (Chhetri et al., 2008). The use of crude oil (which is liquid) in such a systems will be even more efficient than solid fuel. If the crude oil is used directly similar to saw dust electricity generator, the environmental problems associated with fossil fuel use and refining can be minimized increasing the economical efficiency of the fossil fuel use. The CO_2 produced from the direct use of crude oil will be acceptable to plants as it is fresh CO_2 and no catalysts and chemicals are used for processing. Chhetri and Islam (2007a; Chhetri et al., 2008) argued that the CO_2 produced from refined petroleum products may not be acceptable to plants as this CO_2 is contaminated with toxic catalysts and chemicals. Hence, use of direct crude oil will solve the major environmental problems the humanity is facing today.

5.5.6. Use of Gravity Separation Systems

Heavier fractions can be settled out through the density difference method. Various settling tanks in different stages can be designed that allow sufficient time to settle the fractions based on their density. Even though, it will not solve all the problems, some of the environmental problems can be reduced by this method. This will be less costly compared to other processes but more time consuming.

6. ACHIEVING SUSTAINABILITY IN OIL REFINING

Khan and Islam (2006a) developed a green supply chain framework to achieve sustainability in refining operations. The framework proposed supply chain model is developed based on the work of Lakhal and H'Mida (2003) and

Lakhal et al. (2006). It analyses the structure of the supply chain from production, transportation, and distribution to end users. The specific aspects of the model include:

Five zeros of waste or emissions (corresponding to the five circles in the Olympic flag):

a) Zero emissions (air, soil, water, solid waste, hazardous waste)
b) Zero waste of resources (energy, materials, human)
c) Zero waste in activities (administration, production)
d) Zero use of toxics (processes and products)
e) Zero waste in product life cycle (transportation, use, end-of-life)

Green inputs and outputs: The zero waste approach is defended by the Zero-waste Organization (Zero Waste, 2005) using a visionary goal of zero waste to represent the endpoint of "closing-theloop", so that all materials are returned at the end of their life as industrial nutrients, thereby avoiding any degradation of nature. A 100% efficiency of use of all resources – energy, material, and human – is promoted by Zero-waste, working toward a goal of reducing costs, easing demands on scarce resources, and providing greater availability for all. These principles of Zero waste applies to products reduce impact during manufacture, transportation, during use, and at end of life. Such an approach, which is always the norm in Nature, is only beginning to be proposed in the petroleum sector (Bjorndalen et al. 2005) or even in the renewable energy sector (Khan et al. 2005b).

6.1. Zero Emissions (Air, Soil, Water, Solid Waste, Hazardous Waste)

Primary activities in refinery processes are materials transfer and storage, separating hydrocarbons (e.g., distillation), creating hydrocarbons (e.g., cracking/coking, alkylation, and reforming), blending hydrocarbons, and removing impurities (e.g., sulfur removal) and cooling are the major operations to release the solid, liquid and gaseous emission. These emission should be completely controlled so as not to release in the atmosphere, land and water bodies.

6.2. Zero Waste of Resources (Energy, Material, and Human)

The most important resource in the refinery process is energy. Unlike the manufacturing industry, labor costs do not constitute a high percentage of expenses in a refinery. In this continuous process, there is no waste of materials in general. The waste of human resources could be measured by ratios of accident (number of work accidents/number of employers) and absenteeism due to illness (number of days lost for illness/ number of work days × number of employees). In the "Olympic" refinery, ratios of accident and absenteeism would be near to zero.

The refining process uses a huge amount of energy. Typically, approximately 2% of the energy contained in crude oil is used for distillation. The "advanced" or "progressive" distillation, involving the use of modern chemical engineering techniques, cuts down the energy required to distil crude oil by 30 to 65%. This technique requires large-scale rebuilding of distillation units to enable separation of crude oil components. This would avoid the ineptness of conventional distillation processes, like heating oil at high temperatures and separating products as they cool. Advanced or progressive distillation may also include more effective heat exchange, further reducing the energy required for distillation. Considering this potential reduction of energy consumption, it can be estimated that an "Olympic" refinery would use only 1% contained in the crude.

6.3. Zero Waste in Administration Activities

Considerable cost savings could be achieved by using resource-efficient products and good environmental practices. Good practices should target energy-efficiency, waste reduction, water conservation, and other resource-efficient practices for the environment. By taking advantage of these practices, refineries can avoid resource waste and save money. An "Olympic" refinery should have a list of good practices and encourage employers to respect them. For example, one workstation (computer and monitor) left running after business hours, causes power plants to emit nearly one ton of CO_2 per year. That emission could be cut by 80% if the workstation is switched off at night and set to "sleep mode" during idle periods in the day. If every computer and monitor in the United States was turned off at night, the nation could shut down 8 large power stations and avoid emitting 7 million tons of CO_2 every year (Nichols et al. 2001).

6.4. Zero Use of Toxics (Processes and Products)

A number of procedures are used to turn heavier components of crude oil into lighter and more useful hydrocarbons. These processes use catalysts or materials that help chemical reactions without being used up themselves. Refinery catalysts are generally toxic and must be replaced or regenerated after repeated use, turning used catalysts into a waste source. The refining process uses either sulfuric acid or hydrofluoric acid as catalysts to transform propylene, butylenes, and/or isobutane into alkylation products or alkylate. Vast quantities of sulfuric acid are required for this process. Hydrofluoric acid (HF), also known as hydrogen fluoride, is extremely toxic and can be lethal. Using catalysts with fewer toxic materials significantly reduces pollution. Eventually, organic acids and enzymes, instead of catalysts must be considered. Thermal degradation and slow reaction rates are often considered to be the greatest problems of using organic acid and catalysts.

However, recent discoveries have shown that this perception is not justified. There are numerous organic products and enzymes that can withstand high temperatures and many of them induce fast reactions. More importantly, recent developments in biodiesel indicate that the process (Chhetri and Islam, 2008c; Chhetri and Islam, 2008b) itself can be modified in order to eliminate the use of toxic substances. The same principle applies to other materials, for example, corrosion inhibitors, bactericides, etc. Often, toxic chemicals lead to high corrosion vulnerability and even more toxic corrosion inhibitors are required. The whole process spirals down to a very unstable process, which can be eliminated with the new approach (Al-Darbi et al. 2002).

6.5. Zero Waste in Product Life Cycle (Transportation, Use, and End-of-Life)

The complex array of pipes, valves, pumps, compressors, and storage tanks at refineries are potential sources of leaks into air, land, and water. If they are not contained, liquids can leak from transfer and storage equipment and contaminate soil, surface water, and groundwater. This explains why, according to industrial data, approximately 85% of monitored refineries have confirmed groundwater contamination as a result of leaks and transfer spills. To prevent the risks associated with transportation of sulfuric acid and on-site accidents associated with the use of hydrofluoric acid, refineries can use a solid acid catalyst that has recently proven effective for refinery alkylation. A solid acid catalyst for alkylation is much less able than HF to disperse into the environment in a short time frame. Changing to this method would promote inherent safety at a refinery, rather than merely improving accident mitigation and response. An "Olympic" green refinery supply chain should have storage tanks and pipes above ground to prevent groundwater contamination. There is room for improving the efficiency of these tanks with natural additives. Frequently, the addition of synthetic materials makes an otherwise sustainable process unstable. Advances in using natural materials for improving material quality have been made by Saeed et al. (2003).

REFERENCES

Al-Darbi, M., Muntasser, Z., Tango, M., and Islam, M.R. 2002.Control of microbial corrosion using coatings and natural additives. *Energy Sources*, 24(11):1009–18.

Akhtar, J. 2002. Numerical and Experimental Modelling of Contaminant Transport in Ground Water and of Sour Gas Removal from Natural Gas Streams. Chalmers University of Technology, Goteborg, Sweden.

Ahn, J.S. and Lee, K.H., 1986. Studies on the volatile aroma components of edible mushroom (*Tricholoma matsutake*) of Korea. *Journal of the Korean Society of Food and Nutrition*, 15:253–257 [cited in BUA, 1994].

Astrita, G., Savage, D., and Bisio, A. (1983) *Gas Treating with Chemical Solvents.* John Wiley and Sons, New York.

Blomstrom, D.C. and Beyer, E.M. Jr., 1980. Plants metabolise ethylene to ethylene glycol. *Nature*, 283:66–68.

Bjorndalen, N., Mustafiz, S., and Islam, M.R. 2005. No-flare design: converting waste to value addition. *Energy Sources*, 27(4), 371–80.

Basu, A., J. Akhter, Rahman, M.H., and Islam, M.R., 2004, "A review of separation of gases using membrane systems", *J. Pet. Sci. Tech.*, vol. 22, no. 9-10, 1343-1368.

Beyer, K.H., Jr., Bergfeld, W.F., Berndt, W.O., Boutwell, R.K., Carlton, W.W., Hoffmann, D.K., and Schroeter, A.L. (1983) Final report on the safety assessment of triethanolamine, diethanolamine, and monoethanolamine. *J. Am. Coll. Toxicol.* 2, 183-235.

Currie, D.R. and Isaacs, L.R., 2005. Impact of exploratory offshore drilling on benthic communities in the Minerva gas field, Port Campbell, Australia, *Marine Environmental Research*, Vol. 59: 217–233.

Chhetri, A.B., Khan, M.I. and Islam, M.R., 2008. A Novel Sustainably Developed Cooking Stove. *J. Nat. Sci. and Sust. Tech.* 1(4):xxx-xxx.

Chhetri, A.B. and Islam, M.R., 2007a. Reversing Global Warming. *J. Nat. Sci. and Sust. Tech.* 1(1):79-114.

Chhetri, A.B. and Islam, M.R., 2008a. Pathway Analysis of Crude and Refined Oil and Gas. *energies*, accepted.

Chhetri, A.B. and Islam, M.R., 2008b. *Inherently Sustainable Technology Development.* Nova Science Publishers, New York, pp 452.

Chhetri, A.B. and Islam, M.R., 2007a. Scientific Characterization of Energy Sources. *J. Nat. Sci. and Sust. Tech.1(3):359-356.*

Chhetri, A. B. and Islam, M.R., 2008a. Problems Associated with Conventional Natural Gas Processing and Some Innovative. *Energy Sources, 26(13):xxx-xxx.*

Chhetri, A. B. and Islam, M.R., 2008c. Towards Producing Truly Green Biodiesel. *Energy Sources,* 30 (8):754-764.

Chakma, A., 1999. Formulated solvents: new opportunities for energy efficient separation of acid gases. *Energy Sources* 21(1-2).

CEF Consultants Ltd, 1998. Exploring for Offshore Oil and Gas (Nov). No. 2 of Paper

Series on *Energy and the Offshore*. Halifax, NS, <http://www.cefconsultants.ns.ca/2explore.pdf> [Accessed on May 18, 2005].

CERHR, 2003. NTP-CERHR, expert panel report on reproductive and developmental toxicity of propylene glycol. National Toxicology Program U.S. Department of Health and Human Services. NTP-CERHR-PG-03.

Calle, S., Klaba, L., Thomas, D., Perrin, L., and Dufaud, O., 2005, Influence of the size distribution and concentration on wood dust explosion: Experiments and reaction modeling, *Powder Technology*, Vol. 157 (1-3), September, pp. 144-148.

Davis, R.A., Thomson, D.H., Malme, C.I., and Malme, C.I. 1998. *Environmental Assessment of Seismic Explorations.* Canada/Nova Offshore Petroleum Board, Halifax, NS, Canada.

Diviacco, P. 2005. An open source, web-based, simple solution for seismic data dissemination and collaborative research. *Commuters and Geosciences*, 31, 599–605.

De Groot, S. J. D. 1996. Quantitative assessment of the development of the offshore oil and gas industry in the North Sea, *ICES Journal of Marine Science*, vol. 53: 1045-1050.

Deakin, S. and Konzelmann, S.J., 2004. Learning from Enron, *Corporate Governance*, Vol 12, No. 2: 134-142.

Dincer, I. and Rosen, M.A. 2005. Thermodynamic aspects of renewable and sustainable development. *Renewable and Sustainable Energy Reviews*, 9, 169–89.

Environmental Defense, 2004. <http://www.environmentaldefense.org/> (Accessed on June 2, 2006).

EPA, 2000. Development Document for Final Effluent Limitations Guidelines and Standards for Synthetic-Based Drilling Fluids and other Non-Aqueous Drilling Fluids in the Oil and Gas Extraction Point Source Category. EPA-821-B-00-013, U.S. Environmental Protection Agency, Office of Water, Washington, DC 20460, December, <http://www.epa.gov/waterscience/guide/sbf/fi nal/eng.html>.

Fischer, A. and Hahn, C., 2005. Biotic and abiotic degradation behaviour of ethylene glycol monomethyl ether (EGME). *Water Research*. Vol.39:2002–2007.

Fernandes, M.B. and Brooks, P., 2003, Characterization of carbonaceous combustion residues: II. Nonpolar organic compounds, *Chemosphere,* Volume 53, Issue 5, November, pages 447-458.

Gunal, G.O. and Islam, M.R., 2000. Alteration of asphaltic crude rheology with electromagnetic and ultrasonic irradiation", *J. Pet. Sci. Eng.*, vol. 26(1-4), 263-272.

Hull, T.R., Quinn, R.E., Areri, I.G., Purser, D.A., 2002. Combustion toxicity of fire retarded EVA. *Polymer Degradation and Stability* 77:235–242.

Holdway, D.A., 2002. The Acute and Chronic Effects of Wastes Associated with Offshore Oil and Gas Production on Temperature and Tropical Marine Ecological Process. *Marine Pollution Bulletin*, 44: 185-203.

Himpsel, F., 2007. Condensed-Matter and Material Physics: The Science of the World Around Us. The National Academies Press, ISBN-13: 978-0-309-10965-9, pp 224.

Henda, R., Herman, A., Gedye, R., and Islam, M.R. (2005) Microwave enhanced recovery of nickel-copper ore: communication and floatability aspects. *Journal of Microwave Power and Electromagnetic Energy*, 40(1), 7–16.

Islam, M.R., 2004. Unraveling the mysteries of chaos and change: knowledge-based technology development", *EEC Innovation*, vol. 2, no. 2 and 3, 45-87.

Islam, M.R., 2003, Adding Value to Atlantic Canada's Offshore Industry, presented at the 3rd Atlantic Canada Deepwater Workshop, Halifax, N.S., April 9-10, 2003.

Islam, M.R., Shapiro, R., and Zatzman, G.M., "Energy Crunch: What more lies ahead?" The Dialogue: Global Dialogue on Natural Resources, Center for International and Strategic Studies, Washington DC, April 3-4, 2006.

Jepson, P.D., Arbelo, M., Deaville, R., Patterson, I.A.P., Castro, P., Baker, J.R., Degollada, E., Ross, H.M., Herráez, P., Pocknell, A.M., Rodríguez, F., Howie, F.E., Espinosa, A., Reid, R.J., Jaber, J.R., Martin, J., Cunningham, A.A., and Fernández, A. (2003) Gas-bubble lesions in stranded cetaceans. Was sonar responsible for a spate of whale deaths after an Atlantic military exercise? *Nature*, 425, 575–6.

Khan, M.M., Prior, D. and Islam, M.R., 2006b. A Novel, Sustainable Combined Heating/ Cooling/Refrigeration System. *Journal of Nature Science and Sustainable Technology,* vol 1., no. 1: in press.

Khan, M.M., Mills, A., Chaalal, O., and Islam, M.R., 2006c, "Novel Bioabsorbents for the Removal of Heavy Metals from Aqueous Streams", *Proc. 36th Conference on Computer and Industries*, June, Taiwan.

Khan, M. I., Zatzman, G. M., and Islam, M. R., 2005b. A Novel Sustainability Criterion as Applied in Developing Technologies and Management Tools. In *Second International Conference on Sustainable Planning and Development.* Bologna, Italy.

Khan, M.I. and Islam, M.R., 2007a. *A Handbook of Sustainable Petroleum Management and Operations*, 550 pp. (approx.), Gulf Publishing Co., in press.

Khan, M.I. and Islam, M.R., 2003b. Wastes Management in Offshore Oil and Gas: A Major Challenge in Integrated Coastal Zone Management. In: L. G. Luna (ed), *CARICOSTA 2003 - 1st International Conference on Integrated Coastal Zone Management ICZM*, University of Oriente, Santiago du Cuba, May 5-7, 2003.

Khan, M.I. and Islam, M.R. 2006a. Technological analysis and quantitative assessment of oil and gas development on the scotian shelf, Canada. *Int. J. Risk Assessment and Management*, in press.

Khan, M.I. and Islam, M.R. 2005b. Assessing Sustainability of Technological Developments: An Alternative Approach of Selecting Indicators in the Case of Offshore Operations. Proceedings of ASME International, Mechanical Engineering Congress and Exposition, Orlando, Florida, November 5–11.

Khan, M.I., Lakhal, Y.S., Satish, M., and Islam, M.R., 2005. Towards Achieving Sustainability: Application of Green Supply Chain Model in Offshore Oil and Gas Operations. *International Journal Risk Assessment and Management,* vol. 6, in press (accepted for publication in December, 2005).

Khan, M.M. and Islam, M.R., 2006c. A new downhole water-oil separation technique, *J. Pet. Sci. Tech.*, vol. 24, no. 7, 789-805.

Khan, M.I. and Islam, M.R., 2003a. Ecosystem-Based Approaches to Offshore Oil and Gas Operation: An Alternative Environmental Management Technique. SPE Annual Technical Conference and Exhibition, Denver, USA. October 6-8.

Khan, M.I. and Islam, M.R., 2006a. *True Sustainability of Natural Resource Management*, June, 355 pp., Nova Science Publishers, New York.

Khan, M.I. and Islam, M.R., 2007. *A Handbook of Sustainable Petroleum Management and Operations*, 550 pp. (approx.), Gulf Publishing Co., in press.

Khan, M.I. and Islam, M.R., 2005. Assessing Sustainability of Technological Developments: An Alternative Approach of Selecting Indicators in the Case of Offshore Operations. *Proc. ASME International, Mechanical Engineering Congress and Exposition*, Orlando, Florida, Nov 5-11, 2005.

Khan, M. I., Zatzman, G. M., and Islam, M. R., 2005. A Novel Sustainability Criterion as Applied in Developing Technologies and Management Tools. In *Second International Conference on Sustainable Planning and Development.* Bologna, Italy.

Lehman-McKeeman. L.D. and Gamsky, E.A. 1999. Diethanolamine inhibits choline uptake and phosphatidylcholine synthesis in Chinese hamster ovary cells. *Biochem. Biophys. Res. Commun.*262(3):600-604.

Low, N. M. P., Fazio, P. and Guite, P., 1984. Development of light-weight insulating clay products from the clay sawdust-glass system. *Ceramics International* 10(2):59-65.

Lakhal, S.Y. and H'Mida, S. 2003. A gap analysis for green supply chain benchmarking, in *32th International Conference on Computers and Industrial Engineering*, Volume 1, August 11–13, Ireland, pp. 44–9.

Lakhal, S.L., Khan, M.I., and Islam, M.R. (2006a) A framework for a green decommissioning of an offshore platform, in *Proceedings of 36th CIE Conference on Computer and Industrial Engineering*, July, Taiwan, pp. 4345–56.

Matsuoka, K., Iriyama, Y., Abe, T., Matsuoka, M., Ogumi, Z., 2005. Electro-oxidation of methanol and ethylene glycol on platinum in alkaline solution: Poisoning effects and product analysis. *Electrochimica Acta* Vol.51: 1085–1090 MSDS, 2005. Ethylene glycol material data sheet. www.sciencestuff.com/msds/C1721.html (accessed on Nov 22, 2006).

Mehedi, M.Y., Chhetri, A.B., Ketata, C., and Islam, M.R. 2007a. An Approach for Confl ict Resolution in Oil and GasOperations, *Int. J. Risk Assessment and Management*, in press.

Mehedi, M., Khan, M.I., Ketata, C., and Islam, M.R. (2007b) A risk management model for the valued ecosystem components in offshore operations, *Int. J. Risk Assessment and Management*, in press.

Miralai, S., Khan, M.M. and Islam, M.R. 2006. A novel technique for natural disinfection of contaminated water. *J.of Nat. Sci. and Sust.Tech.* In progress.

Nichols, C., Anderson, S., and Saltzman, D. (2001) A Guide to Greening Your Bottom Line Through a Resource-Efficient Office Environment. City of Portland, Office of Sustainable Development, Portland.

Nallinson, R.G., 2004. Natural Gas Processing and Products. Encyclopedia of Energy Vol. IV, Elsevier Publication. Okhlahama, USA, pp235-247.

Piipari R., Tuppurainen, M., Tuomi, T., Mantyla, L., Henriks-Eckerman, M.L.,Keskinen, H., and Nordman, H. 1998. Diethanolamine-induced occupational asthma, a case report. Clin. Exp.*Allergy* 28(3):358-362.

Patin, S.1999. Environmental Impact of the Offshore Oil and Gas Industry. Eco Monitor Publishing, East Northport, New York.

Rao, M.B., Sircar, S., and Golden, T.C. (1992) Gas separation by adsorbent membranes. *US Patent*, 5, 104, 425.

Rao, M.B. and Sircar, S. (1996) Performance and pore characterization of nanoporous carbon membranes for gas separation. *Journal of Membrane Science*, 3(7), 109–18.

Rao, M.B. and Sirkar, S. (1993) Liquid-phase adsorption of bulk ethanol-water mixtures by alumina. *Adsorption Science and Technology*, 10(1–4), 93–104.

Raily, K., 2007. Hemptons. www.hemptons.co.za/Users/seeds/htm (Acessed on October 20, 2007) Sugie, H., Sasaki, C., Hashimoto, C., Takeshita, H., Nagai, T., Nakamura, S., Furukawa, M., Nishikawa, T. and Kurihara, K., 2004. Three cases of sudden death due to butane or propane gas inhalation: analysis of tissues for gas components. *Forensic Science International* 143 (2-3): 211-214.

Saha, S. and Chakma, A.1992. An energy effi cient mixed solvent for the separation of CO_2. *Energy Conversion Management*, 33, 413.

Saeed, N.O., Ajijolaiya, L.O., Al-Darbi, M.M., and Islam, M.R. (2003) Mechanical properties of mortar reinforced with hair fibre. In *Proc. Oil and Gas Symposium, CSCE Annual Conference*, Moncton.

Service, R.F. 2005. Is it time to shoot for the sun? *Science*, 309: 549–551.

Sweis, F.K., 2004, The effect of admixed material on the flaming and smouldering combustion of dust layers, *Journal of Loss Prevention in the Process Industries*, vol. 17 (6), pp. 505-508.

Toninello, A., Pietrangeli P., De Marchi, U. Salvi, M. and Mondov, B., 2006. Amine Oxidases in Apoptosis and Cancer. *Biochimica et Biophysica Acta* 1765:1 – 13.

United Kingdom Offshore Operations Association: Atmospheric Emission (UK OOA). Available http://www.ukooa.co.uk/issues/1999report/ enviro99_atmospheric.htm, March 11, 2003.

Veil, J.A., 2002. Drilling Waste Management: past, present and future. Annual Technical Conference and Exhibition, San Antonio, Texas, 29 Septermber-2 October, 2002. SPE paper no. 77388.

Venkataraman, C., Joshi, P., Sethi, V., Kohli, S., and Ravi, M.R., 2004, *Aerosol Science and Technology*, vol. 38, no. 1, 50-61.

Wenger, L.M., Davis, C.L., Evensen, J.M., Gormly, J.R., and Mankiewicz, P.J., 2004. Impact of modern deepwater drilling and testing fluids on geochemical evaluations, *Organic Geochemistry*, Vol. 35: 1527–1536.

Wasiuddin, N.M., N. Ali, M.R. Islam, 2002b. Use of Offshore Drilling Waste in Hot Mix Asphalt (HMA) Concrete as Aggregate Replacement", paper no. EE 29168, ETCE '02, Feb. 4-6, 2002, Houston, Texas.

Website 1. Balanced Solution.com. Moisture Properties of Plaster and Stucco for Strawbale Buildings www.ecobuildnetwork.org/pdfs/Straube_ Moisture_Tests.pdf (accessed on 8th Aug, 20).

Website 2: www.hemptons.co.za/Users/seeds/htm (accessed on December 15, 2006).

Website 3: http://hyperphysics.phy-astr.gsu.edu/hbase/ems3.html (accessed on November 5, 2006)

Zatzman, G.M, 2007. The Honey-Sugar-Saccharin-Aspartame or HSSA Syndrome: A Note. *J. of Nat. Sci. and Sust. Tech 1*(3). In press.

Zero-Waste, 2005. *The Case for Zero Waste.* <http://www.zerowaste.org/> [Accessed on August 12, 2006].

In: Sustainable Petroleum Operations
Editor: Rafiq Islam
ISBN 978-1-61324-157-8

Chapter 2

A FRAMEWORK FOR SUSTAINABLE PETROLEUM OPERATIONS

M. Ibrahim Khan,[1] M. Satish,[2] M. R. Islam,[2] and S. Y. Lakhal[3]
[1]Institute for Sustainable Energy, Environment, and Economy
ES602-2500 University Dr, NW, University of Calgary,
Calgary, AB, Canada
Email: Ibrahim.Khan@ucalgary.ca
[2]Faculty of Engineering, Dalhousie University-Sexton Campus
D510-1360 Barrington Street, Halifax, NS, B3J 2X4, Canada
[3]Faculty of Business Administration
University of Moncton, Moncton, NB, Canada

ABSTRACT

Oil and gas are considered the lifeblood of modern economic society. However, the exploration and development of oil and gas have recently been found to be environmentally unsustainable. Each phase of hydrocarbon extraction, especially in offshore operations, generates numerous gaseous, liquid, and solid wastes. This paper examines the present status of Offshore Oil and Gas Operations (OOGOs) with respect to waste generation, resource abuse, and toxic compounds usages in its lifecycle. This research analyzes the sustainability of offshore oil and gas operations using the concept of 'green supply chain'. It studies environmental dimensions, company policy, operations, and technology in the context of offshore oil and gas development in the Scotian shelf of Canada. To achieve sustainability in OOGOs, an 'Olympic' green supply chain model has been applied to achieve zero emissions, zero waste of resources, zero waste in activities, zero use of toxics, and zero waste in product life-cycle. However, it has been found that there is no technology available to achieve the objectives of the 'Olympic' supply chain model. Finally, this paper discusses the environmental and technological problems of OOGOs and provides guidelines to achieve overall sustainability in oil company activities.

Keywords: Sustainability in petroleum sectors, offshore oil and gas; Green supply chain; Zero-discharges; Marine environment.

1. Introduction

The oil and gas development industry is one of the largest global economic sectors. Currently, the industry is projected to spend $100 billion on deepwater oil and gas fields, pipelines, drilling rigs, and production platforms around the world. The volume of expenditure is increasing, because this sector is moving toward ultra-deepwater offshore environments (Wenger et al., 2004; Pinder, 2001; Khan and Islam, 2003a). Operations in that environment are based on high technology and monitoring. The large scale investment at stake entails higher risks in terms of economic success and environmental wellbeing and safety.

The success of a high risk offshore oil and gas operation depends on the use of sustainable technology (Khan et al., 2005; Khan and Islam, 2005b) and on the improvement of sustainability in the operating company. Similar to other industrial sectors, the present hydrocarbon development process, which produces oil and gas, is developed for reasons of economic success. Environmental well-being is never considered in the conventional technology development process. As a result, existing best available technology has ended up producing numerous gaseous, liquid, and solid wastes and pollutants (Khan and Islam, 2003b; Holdway, 2002; Veil, 2002; EPA, 2000), none of which has been completely remedied. Therefore, it is believed that OOGOs have negative impacts on habitat, wildlife, fisheries, and biodiversity (Wenger et al., 2004; Holdway, 2002; Khan and Islam, 2003a; Currie and Isaacs, 2005; Schroeder and Love, 2004).

Existing management regimes try to solve the above mentioned problems through developing waste management guidelines, such as the Canada-Nova

Scotia Offshore Petroleum Board (CNSOPB) guidelines and the Environment Protecting Agency guidelines. For example, the CNSOPB guidelines allow the release of produced water containing 10mg/L oil or less (OWTG, 2002). However, other dangerous components of produced water, such as radioactive materials, are not taken into account. These guidelines cannot solve the problem completely. Moreover, they give the operator/industry the legal right to pollute the environment, as long as a certain concentration level is not detected. These problems might be overcome by developing new technologies which guarantee sustainability. Recently, Khan et al. (2005) and Khan and Islam (2005b) introduced a new approach by means of which it is possible to develop a truly sustainable technology. Under these approaches, the temporal factor is considered the prime indicator in sustainable technology development.

In addition to the use of technologies that are unsustainable, the corporate management process itself is based on a structure that resists sustainability. Generally, corporate policy is oriented towards gaining monitory benefits in dishonest, unethical, and improper ways, a fact that became evident in the aftermath of the fall of the world energy giant Enron, in December 2001 (Deakin and Konzelmann, 2004). Post-Enron events, including the crisis that afflicted WorldCom, indicate that practically all corporate structures are based on the Enron model (Zatzman and Islam, 2005). Moreover, management strategies in the oil and gas sector are not well suited to natural resource management. Community-based management is found to be a more suitable approach in natural resource management, including renewable energy development (Khan et al., 2007). In this process, stakeholders, or a community, play a role in a participatory management system and benefit from tangible and intangible values. Present OOGOs management regimes operate in centrally-controlled and un-integrated ways. All their activities, such as planning, central engineering, upstream operations, processing, and supply and transportation groups are undertaken as complete, separated entities (Forest and Oetti, 2003). In this context, it is pertinent to examine the systematic methods of efficiently managing OOGOs supply chains.

Recently, the 'green supply chain' approach has been introduced in supply chain research. A green supply chain provides guidelines to a company to implement the principles of sustainability in the areas of corporate governance, the environment, social responsibility, and economic contribution to the community (Sarkis, 2003). However, the application of green supply chain in the energy development sector is currently at an embryonic stage (Lakhal et al., 2005). Some work has been done on the petroleum supply chain by

Lasschuit and Thijssen (2004), Neiro et al. (2004), and Colella (2005). These works mainly describe business plans that include the supply and marketing of petroleum products, rather than focusing on considerations allied to oil extraction or development, or the technological problems involved in greening production. Lakhal et al. (2005) took the initiative to apply the green supply chain model to the oil refining process of a Canadian company. Few, if any, other works can be cited on this area of research.

This paper examines the present sustainability levels of offshore oil and gas operations in the Scotian shelf of Canada. It introduces an integrated green supply chain model that is structured to achieve sustainability in this sector. The paper proceeds by providing a brief overview of sustainability in OOGOs, followed by details of the methods of waste estimation (Section 2). Section three is devoted to a description of the wastes estimates and proposes measures to incorporate the green supply chain model at this stage. Finally, it analyzes the process by which the green supply chain model leads to environmental sustainability. The concept and model evolved in this study considers data related to Canada's Scotian shelf; however, this methodology should be applicable to any offshore oil and gas development project.

Sustainability in Offshore Operations

Sustainable OOGO development requires a sustainable supply of clean and affordable energy resources that do not cause negative environmental, economic, and social consequences (Dincer and Rosen, 2004 and 2005). The concept of sustainability was brought into the core of social, economic, and environmental debates after the well-known definition of sustainable development devised by the World Commission on Environment and Development (WCED, 1987). However, sustainable development has remained an obscure term due to the absence of explicit guidelines for sustainability. Recently, Khan and Islam (2005b) and Khan et al. (2005) developed an innovative criterion for achieving true sustainability in technological development. This reliable concept can be applied equally effectively to offshore technology development. This kind of new technology should have the potential to be efficient and functional far into the future, in order to ensure true sustainability. Sustainable development is seen as having four elements - economic, social, environmental, and technological. Delivery is the overarching imperative that drives both implementation and further strategy development (Khan and Islam, 2005b).

2. METHODOLOGY

For the purposes of this research, information about the lifecycle of OOGOs and its different activities was collected through review of government documents and reports, and published papers, such as CNSOPB activities reports (CNSOPB, 2005), Khan and Islam (2003a and 2007), and EPA (2000). Severe limitations were encountered in finding information about quantitative data, such as amount of waste generation, amount of emissions, and use of toxic compounds. To overcome this restriction, different types of waste generation were estimated following the methodology of EPA (2000).

Total amounts of drilling wastes, such as drilling cuttings, drilling fluids, and releases of oils were estimated for the four different types of wells used for exploratory and development drilling in shallow and deep waters. Drilling wastes have consequently been estimated using formulae.1 to 8 specified below (source: EPA, 2000).

The dry drill cuttings volume is estimated based on the Equation 1. In this estimations, the dry drilling cuttings are equivalent to gage hole volume plus washout.

Drilling hole volume (ft^3) = {length (ft) × π [diameter (ft)/2]2} × (1 + washout fraction of 0.075)... (1)

drill cuttings (bbls) = hole volume (ft3) x 0.1781 bbls/ft3 (2)

drill cuttings (lbs) = drill cuttings (bbls) x 910 lbs/bbl (3)

Waste Components are estimated following Equations 4 and 5. The units are estimated in lbs. The algebraic calculation of lbs of waste components in the given drilled

Total Wastes (TW) = (base fluid) + (water) + (barite) + (drill cuttings) (4)

TW = (RF × TW) + {[RF × (WF/SF)] × TW} + {[RF × (BF/SF)] × TW} + (DF × TW) (5)

where:
TW = total waste (whole drilling fluid + dry cuttings), in lbs

RF = retort weight fraction of synthetic base fluid
WF = water weight fraction from drilling fluid formulation
SF = synthetic base fluid weight fraction from drilling fluid formulation
BF = barite weight fraction from drilling fluid formulation
DF = drill cuttings weight fraction, calculated as follows

$$DF = 1 - \{RF \times [1 + (WF/SF) + (BF/SF)]\} \quad (6)$$

In order to calculate TW, equations (4) and (5) are first used to calculate DF (Equation 6). Then TW is calculated following Equation 7.

$$TW = \text{drill cuttings (lbs)} / DF \quad (7)$$

Input data to estimate the emissions of OOGOs are shown in Table 1. This data has been gathered from different sources such as EPA (2002) and Wenger et al., (2004). In this estimation, 10.2% (wt./wt) standard (baseline) solids control have been taken into account (EPA, 2000).

The whole drilling fluid volume are estimated following Equation 8.

$$\text{whole SBF volume (bbls)} = \text{synthetic base fluid (bbls)} + \text{water (bbls)} + \text{barite (bbls)} \quad (8)$$

The formation oil in whole mud discharged is 0.2% (vol.), which is calculated based on Equation 8.

$$\text{formation oil (bbls)} = 0.002 \times \text{whole SBF volume (bbls)} \quad (9)$$

In the case of green supply chain analysis, the methodology used follows Lakhal et al.,'s (2005) work. In this study, the "Olympic" green supply chain method is used to achieve the following five "zeroes", corresponding to the five circles of the Olympic logo: (i) zero emission (air, water, solid wastes, soil, toxic wastes, hazardous wastes; (ii) zero resource wastes; (iii) zero wastes in activities (energy, materials, human); (iv) zero use of toxics; and (v) zero waste in product life-cycle in case of OOGOs.

Table 1. Characteristics of SBF drilling muds (EPA, 2000)

Waste Characteristics	Value	Waste Characteristics	Value
SBF formulation Synthetic base fluid Barite Water Percent (vol.) formation oil	(by weight) 47% 33% 20% 0.2%	SBF density SBF drilling fluid density Barite density Water Drilling cuttings	280 lbs/bbl 9.65 lbs/ga 1,506lbs/bbl 350 lbs/bbl 910 lbs/bbl
Priority Pollutant Organics	lbs/bbl of SBF	Non-Conventional Organics	lbs/bbl of SBF
Naphthalene Fluorene Phenanthrene Phenol	0.0010024 0.0005468 0.0012968 0.000003528	Alkylated benzenes Alkylated naphthalenes Alkylated fluorenes Alkylated phenanthrenes Alkylated phenols Total biphenyls Total Dibenzothiophenes	0.0056429 0.0530502 0.0063859 0.0080683 0.0000311 0.0104867 0.0004469
Priority Pollutant Metals Cadmium Mercury Antimony Arsenic Beryllium Chromium Copper Lead Nickel Selenium Silver Thallium Zinc	mg/kg Barite 1.1 0.1 5.7 7.1 0.7 240.0 18.7 35.1 13.5 1.1 0.7 1.2 200.5	Non-Conventional Metals Aluminium Barium Iron Tin Titanium	mg/kg Barite 9,069.9 588,000 15,344.3 14.6 87.5
		Conventionals	lbs/bbl of SBF
		Total Oil as SBF Total Oil as formation oil TSS as barite	190.5 0.588 133.7

3. Results and Discussion

3.1. Emission from OOGOs

This research examined the supply chain of offshore oil and gas development in the Scotian shelf of Canada. A typical offshore oil and gas project have five primary phases. These are: seismic exploration, drilling, production, transportation, and decommissioning. The overarching life cycle has been pictorially represented in Figure 1.

The five cylinders represent five different phases of the life cycle. The average operating periods for each phase are 20-30 days, 3-5 yrs, 20-30 yrs, 15-25 yrs, and 4-6 months, tentatively estimated for seismic exploration, drilling, production, transportation, and decommissioning respectively. The left arrows represent the input of chemicals or activities. The right arrows show the qualitative emissions or impacts on the environment. The seismic and drilling activities on the Scotian shelf are shown in Figures 1 and 2. A detailed report on the Scotian shelf can be found in the study of Khan and Islam (2007).

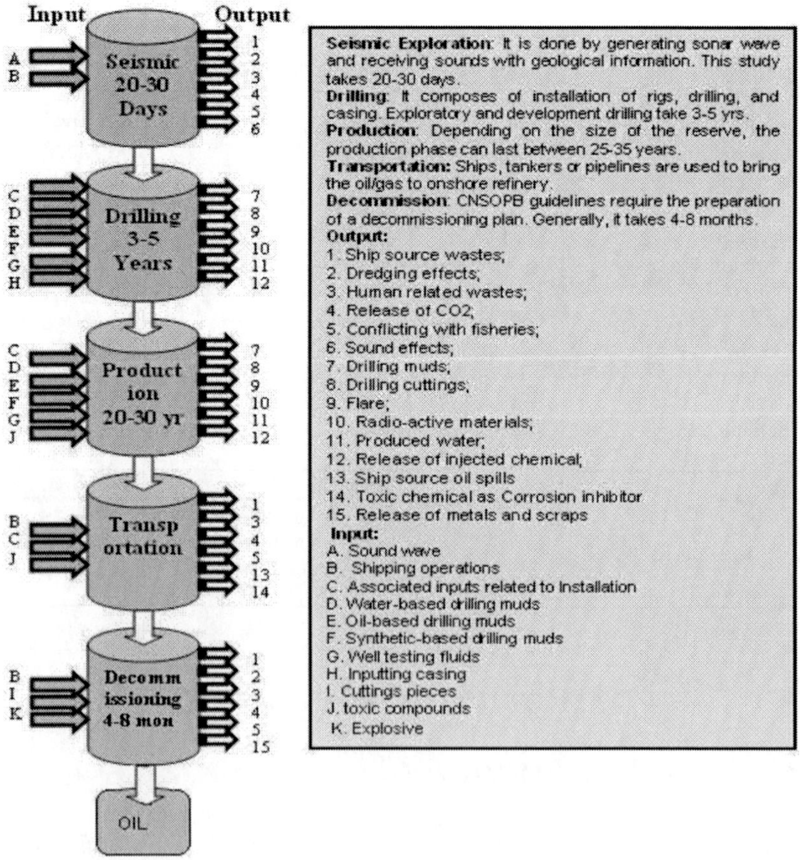

Figure 1. Supply chain of offshore oil and gas development.

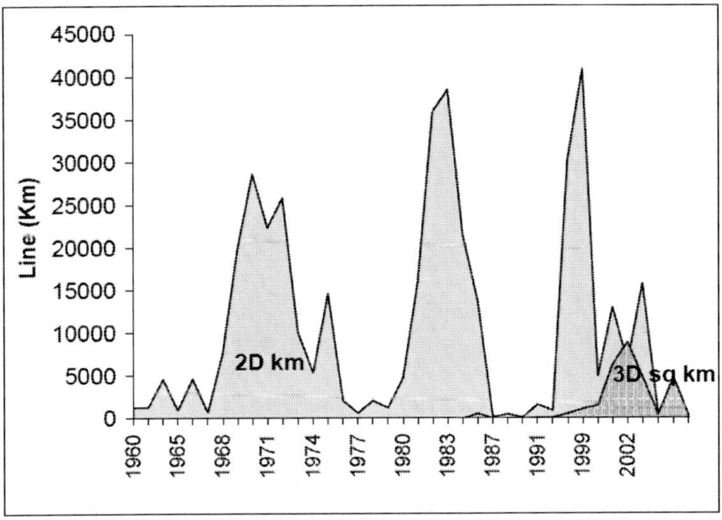

Figure 2. 2D and 3D seismic exploration in the Scotian shelf, Canada from 1964 to 2004.

In this study, major emissions and related activities have been identified and are shown in Figure 1. From the seismic to the decommissioning phases of OOGOs (Figure 1), the major wastes are the following: ship source wastes, human related wastes, release of CO2, conflict with fisheries, dredging effects, sound effects, drilling mud, drilling cuttings, flare, radio-active materials, produced water, release of injected chemical, and release of metals and scraps. Similar types of wastes from offshore operations have also been reported by Khan and Islam (2003a). It is also reported that generation of wastes and other pollutant emissions is the main contentious issue in oil and gas development (Khan and Islam (2003a, 2003b, 2004, 2005a, 2005b, 2005c). These emissions have negative effects on fisheries, biodiversity, special habitat, wildlife, benthic habitat, as substantiated by Holdway (2002), Schroeder and Love (2004), Khan and Islam (2004), and Wells et al. (1995).

Every phase of the offshore operation lifecycle is accompanied by development and generates wastes. The phases of seismic exploration, drilling, production, transportation, and decommissioning generate 6, 7, 8, 4, and 6 types of wastes respectively (Figure 1). Of all these phases, the drilling phase generates the highest volumes of wastes and toxicity levels. In the wake of this fact, this study focuses on quantitative wastes generation from drilling activities.

Table 2. Estimated major wastes components of shallow and deep water development and exploratory drilling wells

Waste Component	Shallow Water (kgs)		Deep Water (kgs)	
	Development	Exploratory	Development	Exploratory
SBF	30,381	63,666	45,975	102,221
Water	12,928	27,092	19,564	43,498
Barite	21,332	44,702	32,280	71,772
D-Cuttings	233,215	488,719	352,918	784,674
A-Cuttings	297,856	624,178	450,737	1,002,165
W-SBF	64,641	135,460	97,819	217,491
F-fluid	94	196	142	316

Table 3. Estimated priority pollutants from shallow-water and deepwater drilling wells

Priority Pollutant Organics	Shallow Water (kgs)		Deep Water (kgs)	
	Development	Exploratory	Development	Exploratory
Cadmium	2.34599178	4.91621664	3.5501159	7.8933036
Mercury	0.21327198	0.44692879	0.3227378	0.7175731
Antimony	12.1565029	25.4749407	18.396055	40.901664
Arsenic	15.1423106	31.7319437	22.914385	50.947687
Beryllium	1.49290386	3.1285015	2.2591647	5.0230114
Chromium	511.852752	1072.62908	774.57074	1722.1753
Copper	39.8818603	83.5756828	60.35197	134.18616
Lead	74.858465	156.872004	113.28097	251.86814
Nickel	28.7917173	60.335386	43.569604	96.872362
Selenium	2.34599178	4.91621664	3.5501159	7.8933036
Silver	1.49290386	3.1285015	2.2591647	5.0230114
Thallium	2.55926376	5.36314542	3.8728537	8.6108766
Zinc	427.61032	896.092214	647.08931	1438.734
Non-Conventional Metals (kgs)				
Aluminum	19343.5553	40535.9939	29271.997	65083.158
Barium	1254039.24	2627941.26	1897698.3	4219329.5
Iron	32725.0924	68578.0936	49521.858	110106.56
Tin	31.1377091	65.2516026	47.11972	104.76567
Titanium	186.612983	391.062687	282.39558	627.87642

Table 4. Estimated organic compounds in shallow and deepwater development and exploratory wells

Priority Pollutant Organics	Shallow Water (lbs)		Deep Water (lbs)	
	Development	Exploratory	Development	Exploratory
Naphthalene	0.5924184	1.2409712	0.8961456	1.9937736
Fluorene	0.3231588	0.6769384	0.4888392	1.0875852
Phenanthrene	0.7664088	1.6054384	1.1593392	2.5793352
Phenol	0.00208505	0.00436766	0.003154	0.0070172
Non-Conventional Organics		lbs		
Alkylated benzenes	3.3349539	6.9859102	5.0447526	11.223728
Alkylated naphthalenes	31.3526682	65.6761476	47.426879	105.51685
Alkylated fluorenes	3.7740669	7.9057442	5.7089946	12.701555
Alkylated phenanthrenes	4.7683653	9.9885554	7.2130602	16.047849
Alkylated phenols	0.0183801	0.0385018	0.0278034	0.0618579
Total biphenyls	6.1976397	12.9825346	9.3751098	20.858046
Total Dibenzothiophenes	0.2641179	0.5532622	0.3995286	0.8888841

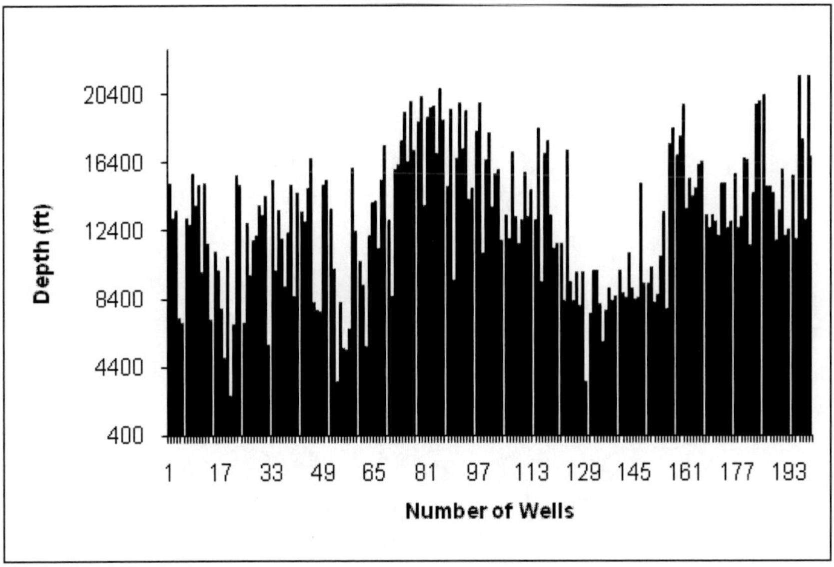

Figure 3. Depths of wells drilled on the Scotian shelf from 1967 to 2003.

Wastes from a typical drilling well come from drilling mud, cuttings, barite, formation fluid, organic compounds (including oil compounds), and heavy metals concentrations. Estimated levels of wastes are shown in Tables 2, 3, and 4. In this study, synthetic-base drilling mud has been considered. Generally, water-base, oil-base, and synthetic-base mud/fluids are used in drilling. The detailed compositions of different types of drilling mud are shown in Figure 4. Due to considerations of toxicity and environmental degradation, synthetic-based mud types are recommended in Europe, USA, and Canada (EPA, 2000; CNSOPB, 2000, and Wenger et al., 2004).

Due to its high toxic levels, OBM is not recommended for use. In case of waste estimation, SBM is favored because it has been widely used (EPA, 2000; Khan and Islam, 2007). The oil retention level of the mud was considered as only 10.2%, which corresponds to conventional uses (EPA, 2000).

On the Scotian shelf, exploratory and development wells have been drilled to different depths (Figure 3). Drilling data of the Scotian shelf from 1964-2004 has been analyzed. For wastes estimations, the wells are divided into shallow water (less than 1000ft) and deep water (higher than1000ft). Exploratory and development drilling has been conducted in both shallow and deepwater. In this study, four categories of wells have separately been estimated and presented in Tables 2, 3, and 4.

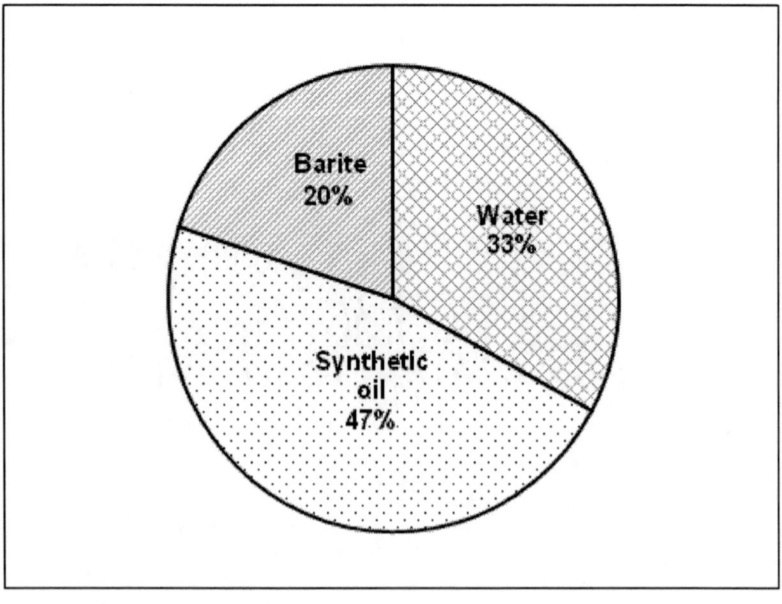

Figure 4. Different composition of drilling mud.

Table 2 shows that the major wastes components are drilling cuttings, SBS, formation oil, water, and barite. Figure 5 shows comparative wastes generation of deep and shallow water during exploratory and development drilling. For all types of wells, exploratory drilling in deepwater generates the maximum amount of wastes. The solid wastes associated with drilling are dry-cuttings (includes 7.5% washout), cuttings and adherent drilling fluid generated from SBF/OBF interval (wet-cuttings), and barite. The solid wastes, including dry-cuttings, wet-cuttings, and barite that are generated from deepwater exploratory drilling are 71,772, 784,674, and 1,002,165 kgs respectively. The estimated amount for the other three types of wells is presented in Table 2. Barite is another solid waste which is generated in the highest volume from the same well (Figure 5).

The total amount of SBF used in shallow-exploratory, shallow-development, deep-exploratory, deep-development drilling are 63666, 30381, 102221, and 45975 kgs per well respectively. SBF is the main cause of concern because it contains purified oil and other substances. Its detailed composition is shown in Figure 4 and Table 1. The detailed result of SBF wastes is shown in Figure 5.

Barite is generated as a solid waste in drilling. It is used as a weight agent in all kinds of drilling fluids mainly composed of priority pollutant and non-pollutant metals. Barite is one of the most harmful pollutants (Wenger et al., 2004; EPA, 2000). The composition of a variety of metals is shown in Table 2. The priority pollutant metals contained in drilling mud are ccadmium, mercury, antimony, arsenic, beryllium, chromium, copper, lead, nickel, selenium, silver, thallium, and zinc. The estimated composition of these heavy metals for exploratory drilling in deepwater is shown in Figure 7.

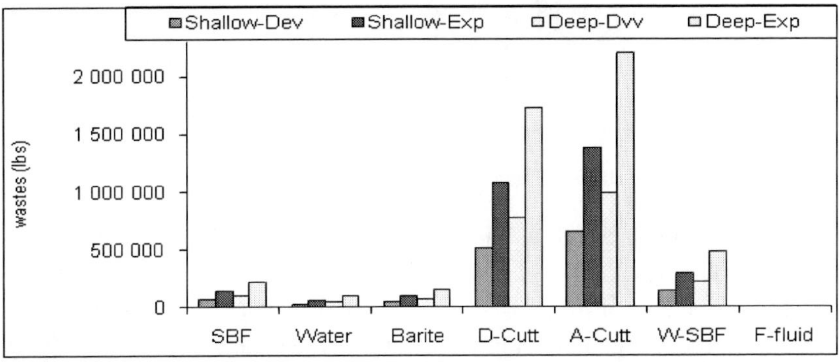

Figure 5. Wastes component of shallow and deepwater during exploratory and development drilling.

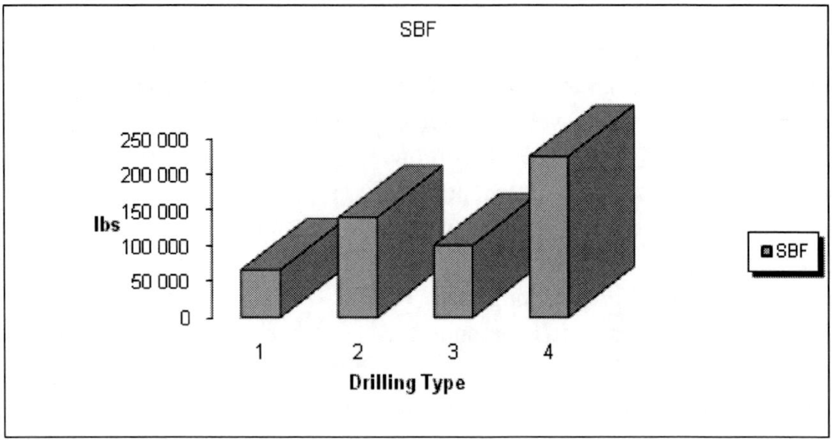

Figure 6. Total amounts of synthetic base fluid (SBF) used in the shallow and deep water wells.

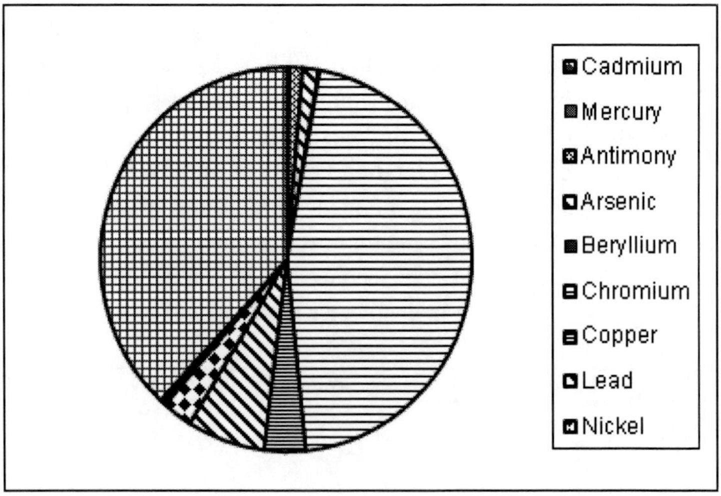

Figure 7. Priority pollutant metals in a deepwater development well.

In this study, it was found that many organic compounds (oil compounds) are emitted with other wastes. The detailed result of organic compounds released from four drilling wells is shown in Table 4. According to EPA (2002), emitted organic compounds are divided into two types, priority pollutants and non-conventional organic compounds (Table 4).

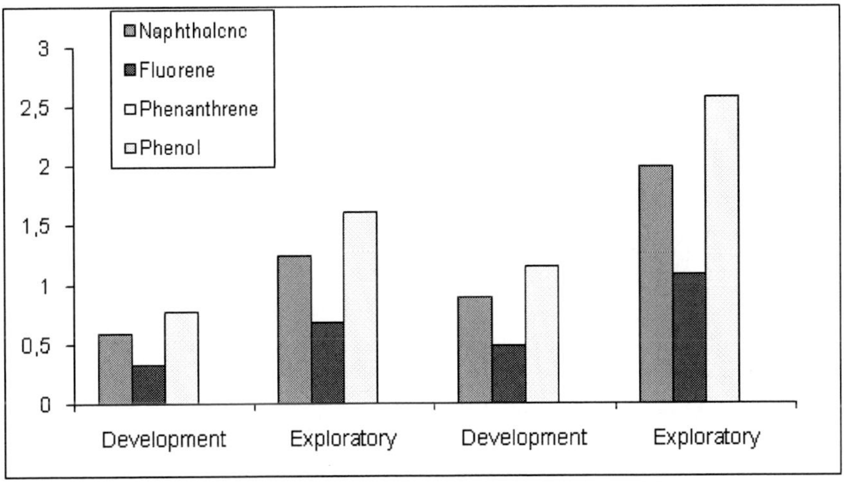

Figure 8. Comparative amounts of Naphthalene, Fluorene, Phenanthrene, and Phenol in shallow and deepwater development and shallow wells.

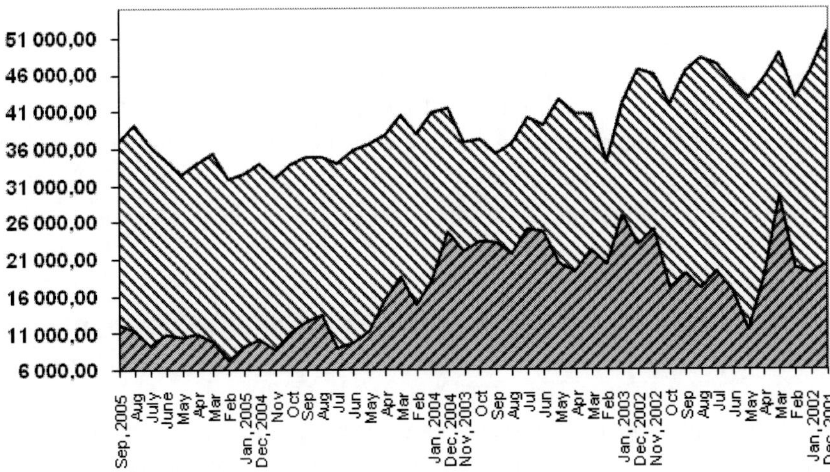

Figure 9. Total amount of gas and water produced in the Sable Island offshore energy project on the Scotian Shelf.

Priority pollutants are naphthalene, fluorene, phenanthrene, and phenol. Figure 8 shows the comparative amount of priority pollutant in the four different wells. Naphthalene was found in the highest amount in all four wells, while phenol had the lowest levels for all wells. Non-conventional organic compounds found were alkylated benzenes, alkylated naphtalenes, alkylated fluorenes, alkylated phenanthrenes, phenols, total biphenys, and total dibenzothiophenes. The estimated composition of priority and non-conventional organic compounds in drilling wastes is presented in Table 4.

During the production period, two major emissions occurred. One was from flaring and other from produced water. Table 5 shows the pollutant types, emission factor, and total amounts emitted from flaring in a singe well on the Scotian shelf. CO_2 constituted the highest amount of pollutants - 1100 tons. After flaring, the next highest emissions resulted from the drilling platform and associated equipments. CNSOPB (2005) estimated the release of CO_2 from a drill ship and drilling project at 22443 and 16161 tons per well respectively for a 90-day operation.

Produced water emits the highest amount of pollutants released from OOGOs (Khan and Islam, 2005a). Figure 9 shows monthly production of produced water, along with gas produced from the Sable Island Offshore Energy Project (SIOEP) and the Scotian shelf from 2001 to 2005. Produced water contains heavy metals, oil, and brine solutions (Khan and Islam, 2005a).

In addition to the above mentioned wastes, huge amounts of waste are produced when decommissioning a project. These wastes are produced by the large amount of materials utilized in the construction of production facilities. For example, the SIOEP project used more than 900,000 meters of pipe, 109 million kg of steel, 53,000 meters of chrome tubing, and 160,000,000 kg of concrete. According to present existing guidelines oil structure has to remove based on the depth of the water (Khan and Islam, 2007). However, following the accurate guidelines still lot of materials are left on the production in sites.

3.2. Application of the Green Supply Chain in OOGOs

The supply chain of offshore oil and gas operations is developed on the basis of their lifecycle activities. Figure 1 illustrated detailed supply chain of OOGOs. Cylinders 1 to 7 indicate the process of operations and finally output of these process in oil. Seismic exploration is at the top level of the chain and decommissioning is at the bottom most level. The intermediate steps include drilling, production, and transportation. The solid arrows on the left-hand side represent input, while the blank arrows on the right-hand side represent emission. A detailed description of Figure 1 has been presented in section 3.1.

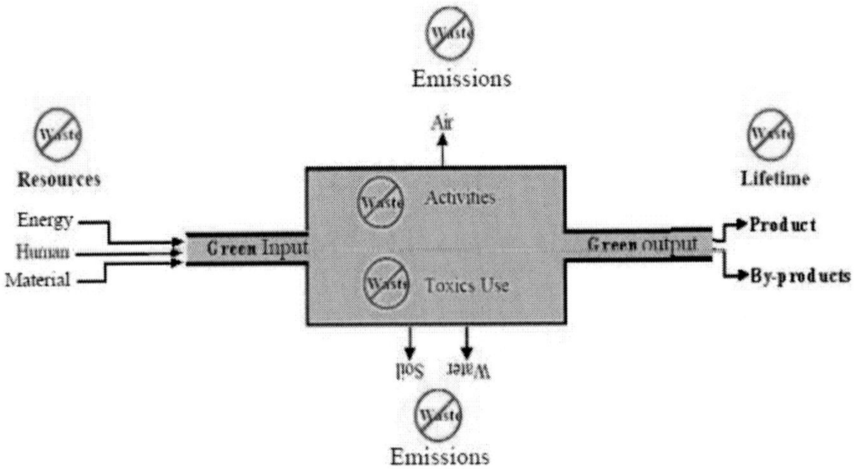

Figure 10. The proposed "Olympic" green supply chain for offshore oil and gas operations (adopted from: Lakhal et al., 2005).

At present levels of operation, activities associated with the OOGOs supply chain emit numerous wastes into the air, water, and sediments (Figure 1). The seismic process emits 10 wastes, while the drilling process is the source of the highest emissions within the chain. Major emissions from OOGOs, especially drilling wastes, are detailed under the discussion of estimates in section 3.1. The main wastes of OOGOs that impact the environment are drilling-waste fluids or muds, drilling-waste solids, produced water, and volatile organic compounds. Moreover, produced water constitutes the highest amount of waste released in the whole supply chain (Figure 9) (Khan and Islam, 2005a and 2007).

Analogous to the "Olympic" logo, the proposed supply chain for OOGOs consists of five zeros (zero-wastes), which are: (i) zero emissions; (ii) zero resource wastes; (iii) zero wastes in activities; (iv) zero use of toxics; and (v) zero waste in product life-cycle. Figure 10 captures the concept of the Olympic supply chain for OOGOs in a representative diagram. According to this model, the system is enabled to input green resources. In Figure 10, energy, human, and materials are shown as green inputs. The rectangular box illustrated as a processing unit shows that toxic compounds cannot be used in the chain. Production and administrative activities can be wasted in regards to inefficiency in processing.

The three arrows at the top of the picture indicate that the chain admits no gaseous, solid, and liquid emissions or discharges. The outlet of the system also shows green products that generate no product lifecycle, in the form of transportation, use, and end-of-use of the product. This supply chain model can be used to achieve sustainability in OOGOs. This is because such an approach is always the norm in nature. This is a new approach for use in the petroleum (Bjorndalen et al., 2005) and renewable energy sectors (Khan et al., 2005 and 2007).

3.2.1. Zero Emission of Air, Water, Solid Wastes, Soil, Toxic Wastes, and Hazardous Wastes

It is evident from the discussion in section 3.1 that OOGOs generate huge amounts of solid, liquid, and gaseous wastes which are released into the air, water, and sediments. Figures 1, 4, 5, 6, 7, and 8 and Tables 2 to 5 clearly illustrate how every process in OOGOs emits wastes. In the proposed "Olympic" green supply chain model (Figure 11), it has been demonstrated that it is possible to release no wastes into air, water, and soil, a goal that present OOGOs have not been able to achieve. As a consequence, current offshore operations have negative impacts on the environment (Holdway, 2002; Khan and Islam, 2003a, 2004, 2005a, 2005b and 2005c; Colella et al., 2005; Currie and Isaacs, 2005).

To achieve sustainability in the OOGOs sector, it is imperative to stop all emissions that the sector is currently producing. The challenge facing us is to devise ways to stop the emissions substantially so as to bring them down to zero-levels in OOGOs. Following a general evaluation of currently available technology, it has been concluded that no technology is currently available that could completely eliminate emissions (Khan and Islam, 2008). Therefore, the need to develop sustainable technology is extremely urgent, as proposed by Khan et al. (2005).

Some technologies do exist that reduce discharges considerably. For example, solids wastes (drill cuttings and barites) can be reduced significantly by adding extra shale screen or centrifuger during the drilling process. Re-injection or onshore disposal is also considered as an effective way to achieve zero emission of solids, liquids (produced water), and gaseous (CO_2) emissions (Malik and Islam, 2000). In addition to injection and implementation of new cleansing processes, other technical and operational modifications at installation sites can contribute to lowering the levels of some specific wastes.

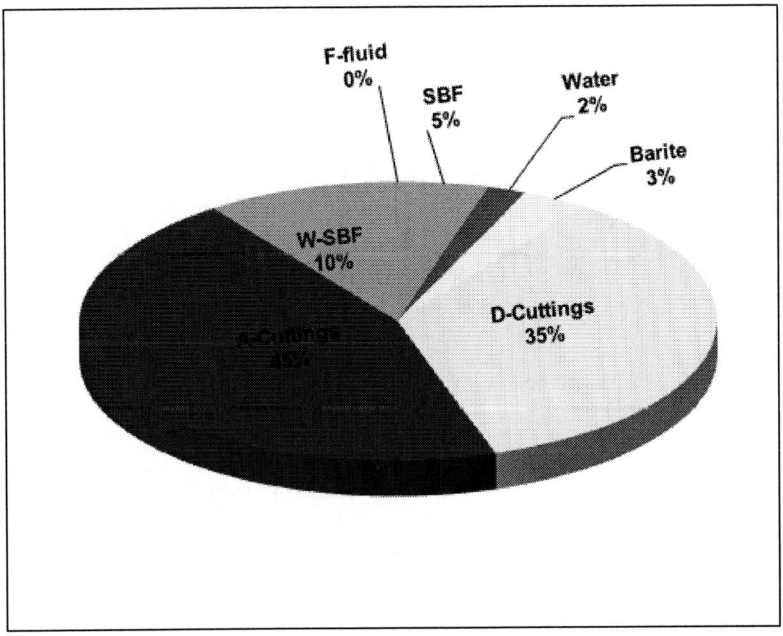

Figure 11. Total drilling wastes from a deepwater development well.

3.2.2. Zero Waste of Resources (Energy, Materials, and Human)

The proposed supply chain model (Figure 10) requires zero waste generation from input materials, involvement of humans, and use of energy. Present OOGOs lose many resources in these areas through flaring, blow-outs, and spills (Table 5). Drilling fluids are generally re-used but large amounts are washed out with drilling cuttings. Table 2 represents a single development well in deepwater that lost 2,17,491 kgs of SBF and 71,772 kgs of barite. This kind of loss can be significantly reduced by developing new eco-friendly technology. To reduce drill cutting wastes, more shale screen and centrifuger may be used. Use of more screens sometimes turns the small sized particles into finer sizes, making it hard to separate and reduce SBF from the fine solid particles.

An offshore operation is not a high, energy-intensive project like projects of other manufacturing industries. However, there is still a lot of room to improve energy use, such as reduction of drilling periods. Drilling a well takes a couple of months to a year (Figure 1). This time can be significantly reduced by employing better technology. Besides drilling time, durations of seismic exploration, production, and decommission can also be reduced by use of improved technology.

Fatal accidents are reported very often In OOGOs. These are frequently associated either with lifting incidents on decks or with drilling activity. Table 6 shows the fatality and injury figures per 100,000 employees. According to Powell (2004), although no fatalities were reported between April 2002 and March 2003, three deaths have occurred since. Recently, the oil firm Shell was fined £900,000 following the death of two workers on the Brent Bravo platform in the North Sea in September 2003 (BBC, 2005).

Table 5. Emissions associated with flaring in one well (CEF, 2005)

Pollutant	Emission Factor (kg/1000 m^3)	Total Emissions (tonnes)
NOx	1.1	0.616
SO2	0.27 *(H_2S in pppmv/100	0.00756
PM2.5	0.61	0.342
Benzene	0.0025	0.0014
Total PAHs	0.000048	0.000027
CO2 (CO_2E)	1913	1100
CH4 (CO_2E^2)	0.04*23	0.515
N2O (CO_2E^2)	0.04*296	663

Table 6. Total fatalities/injuries among 100,000 employees (source: Powell, 2004)

Category	1999/2000	2000/2001	2001/2002	2002/2003p
Fatalities	2 (10.5)	3 (12.9)	3 (12.9)	0 (0)
Major injuries	53 (227.2)	53 (227.2)	47 (202.5)	64 (287.5)
Over 3-day injuries	193 (1015.8)	177 (758.7)	187 (805.8)	118 (530.0)

3.2.3. Zero Wastes in Activities (Administration, Production)

The proposed "Olympic" green supply chain demands no activity wastes in regard to production and administration (Figure 10). It is vital for OOGOs to achieve this goal because these operations involve high expenditure. Reduction in activity means reduction in monetary expenditure. Only big companies are involved in OOGOs and their current management systems are centrally controlled. However, their operations are in remote areas and sometimes field geologists or engineers cannot take *in situ* decisions in response to the particular behavior of a well. It has been reported by Lojan (2005) that drilling activities had to be stopped in an Ecuador oil field to await a simple decision from the company head office in China. According to Lojan, this entails lot of revenue loss and is an example of wasteful administrative activities that involve high costs.

An alternative solution to the existing bottom-up approach in OOGOs is the introduction of a community-based management regime. This approach is effective in securing administrative, production, and other environmental benefits (Khan et al., 2005b). Figure 12 indicates the major difference between existing management practices and the proposed community-based management arrangement. At the moment, most of the oil industry still does its planning, engineering, upstream operations, refining, and supply and transportation in a centralized, non-integrated way (Forest and Oetti, 2003), a structure that is opposed to the proposed community-based model.

3.2.4. Zero Use of Toxics

In the present practice of OOGOs, many toxic compounds are used in drilling fluids, well testing chemicals, and corrosion inhibitors.

Table 3 shows the types and amounts of organic pollutants used in the drilling process. Instead of using toxic compounds, natural non-toxic compounds can be used in this process. AlDarbi et al (2002) found that natural vegetable components are highly efficient in controlling pipeline corrosion. They suggested replacing conventional toxic corrosion inhibitors with these natural components. Figure 1 shows the composition of drilling fluids which have many toxic compounds, including carcinogens. Recently, many of these compounds were replaced by vegetable based compounds (EPA, 2000).

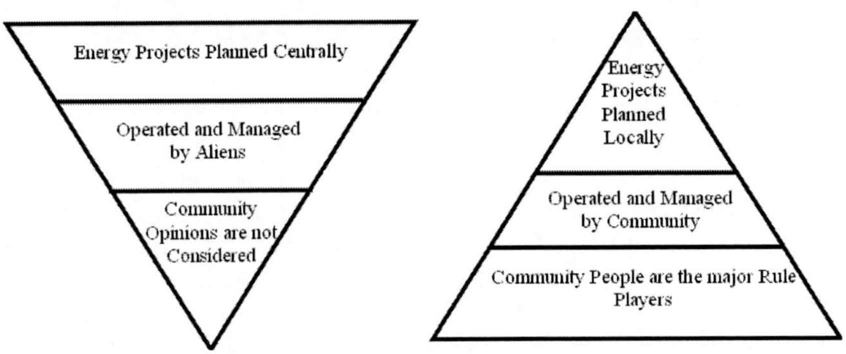

Figure 12. Upside-down, bottom-up approach and community-based energy development approach (Khan et al., 2005b).

3.2.5. Zero Waste in Product Life-Cycle

An "Olympic" green supply chain would stop wastes in the lifecycle, including transportation, usages, and end-of-life of a product. In existing offshore operations, transportation of oil involves many accidents resulting in large amounts of oil spills. It has been reported that on an average from the offshore oil and gas operation 15 million gallons of crude oil release yearly into the marine environment (Khan and Islam, 2003b). There are in total 700 million gallons or oil discharges from different other sources into the sea (Khan and Islam, 2003b and 2004).

Decommissioning a platform also generates large amount of waste products. The production installation is made up of large amounts of material whose dismantling is a big environmental concern. For example, as described in Section 3.1, a single energy project can use more than 900,000 meters of pipe, 109 million kg of steel, 53,000 meters of chrome tubing, and 160,000,000 kgs of concrete. Some innovative technologies have been suggested by Khan and Islam (2003b) to achieve zero waste in the product lifecycle of OOGOs, including rigs to utilize artificial reefs and fish shelter, and platonic utilization of wastes.

3.3. Supporting Sustainability

The proposed "Olympic" supply chain for OOGOs supports the major goals of sustainability through improved technology, economic wellbeing, environmental protection, and social benefits. It was mentioned earlier that no current technology is suitable to completely achieve the zero waste objective, and that there is a need to develop new technology to meet this goal. This new technology must ensure sustainability following the recently developed criteria of sustainability proposed by Khan et al. (2005).

Economic well-being is improved when organizations are able to identify inefficiencies in processes, products, and services and are consequently enabled to find cost-saving solutions for them. For example, waste reduction improves efficiency and lowers costs. Zero waste generation will directly help to reduce regulatory compliance and project costs. By maintaining good environmental standards, a company will acquire a good reputation and greater consumer acceptance, achieving indirect long-term benefits.

Reduction of wastes to zero levels would significantly improve the environmental quality of areas where offshore oil and gas development is conducted. At present, OOGOs are blamed for adverse environmental effects due to the wastes they release (Holdway, 2002; Wells et al., 1995; Khan and Islam, 2003a). Following compliance with the "Olympic" green supply chain model (Figure 10), social wellbeing can be achieved through reliance on the proposed new management technology which would ensure that oil and gas resources are available to all. In addition, a more complete use of "wastes" will create jobs, such as logistics and reprocessing activities. For example, the development of Titanium Corporation has turned out to be a profitable venture as the corporation produces valuable products from oil industry wastes (Eliot, 2005).

Conclusions

This paper has examined the present status of offshore oil and gas operations and its relation to waste generation, resource wastage, and toxic compounds usage in its lifecycle. This study confirmed that every phase of the OOGOs lifecycle generates enormous amounts of gaseous, liquid, and solid wastes. In terms of emissions and use of toxic components, drilling was identified as the most critical waste-producing phase. In quantitative terms, the study estimated wastes generated by exploratory and development drilling in different water depths. The study also identified inefficient resource utilization in the forms of energy, human, and materials, and toxic compound usage in different processes of the production lifecycle. The study has proposed the model of the "Olympic" green supply chain, a new strategy to achieve sustainability in offshore operations.

The main objective of the "Olympic" green supply chain is to achieve zeros emissions, zero resources, zero production activities, zero use of toxic compounds, and zero wastes in product lifecycle in the functioning of OOGOs. This study has concluded that a great deal still needs to be done to achieve the desired goal of "five zeros' in OOGOs. For example, there is no available technology to achieve zero-discharges in drilling and produced waters. This is because existing technologies were developed only to achieve the economic goals of projects and disregarded aspects of environmental wellbeing. Therefore, the study strongly recommends that OOGOs introduce innovative technologies to achieve the proposed "Olympic" supply chain goals. Implementation of the proposed supply chain model will greatly help oil companies to develop cost efficient and take steps towards realizing the global desire for long-term sustainable business practices and processes.

REFERENCES

AlDarbi, M., Muntasser, Z., Tango, M., and Islam, M.R., 2002, "Control of Microbial Corrosion Using Coatings and Natural Additives", *Energy Sources*, vol. 24, no. 11, 1009-1018.

BBC, 2005. "Oil worker deaths inquiry begins", BBC online news. <http://news.bbc.co.uk/>, [Accessed: Oct 30, 2005].

Bjorndalen, N., Mustafiz, S., M.H. Rahman and Islam, M. R. (2005). "No-Flare Design: Converting Waste to Value Addition". *Energy Sources*, Vol 27: 371-380.

CEF, 2005. "Strategic environmental assessments of the Misaine Bank area", prepared for Canada-Nova Scotia Offshore Petroleum Board, Halifax, Canada. 136 pp.

CNSOPB, 2005. "Activity Report 2005", Canada-Nova Scotia Offshore Petroleum Board, Halifax, Nova Scotia. November 5, 2004.

Colella, W.G., Jacobson, M.Z., and Golden, D.M., 2005. "Switching to U.S. hydrogen fuel cell vehicle fleet: The resultant changes in emissions, energy use, and greenhouse gases". *Journal of Power Sources*, Vol 150: 150-182.

Currie, D.R. and Isaacs, L.R., 2005. "Impact of exploratory offshore drilling on benthic communities in the Minerva gas field, Port Campbell, Australia", *Marine Environmental Research*, Vol 59: 217–233.

Deakin, S. and Konzelmann, S.J., 2004. "Learning from Enron", *Corporate Governance*, Vol 12, No. 2: 134-142.

Dincer, I. and Rosen, M., 2004. "Exergy as a drive for achieving sustainability". *International Journal of Green Engineering*, Vol 1, No. 1: 1-19.

Dincer, I. and Rosen, M. A., 2005. "Thermodynamic aspects of renewable and sustainable development", *Renewable and Sustainable Energy Reviews*, Vol 9: 169-189.

Elliott, G., 2005. Personal communication, Executive Chairman, Titanium Corporation, Toronto, Ontario, Canada.

EPA, 2000."Deovelopment document for final effluent limitations guidelines and standards for synthetic-based drilling fluids and other non-aqueous drilling fluids in the oil and gas extraction point source category". United States Environmental Protection Agency. Office of Water, Washington DC 20460, EPA-821-B-00-013, December 2000.

Forest, J and Oetti, M., 2003. "Rigorous simulation supports accurated refinery decisions". *In* I. E. Grossman and C.M. McDonald (Eds)

Proceedings of fortune international conference on foundations of computers-aided process operations, Coral Springs, CAChE: 273-280.

Holdway, D.A.,2002. "The acute and chronic effects of wastes associated with offshore oil and gas production on temperate and tropical marine ecological processes". *Marine Pollution Bulletin*, Vol 44: 185–203.

Islam, M.R., 2005. "Knowledge-Based Technologies For The Information Age", JICEC05-Keynote speech, Jordan International Chemical Engineering Conference V, 12-14 September 2005, Amman, Jordan.

Khan, M.I., and Islam, M.R., 2003a. "Wastes Management in Offshore Oil and Gas: A Major Challenge in Integrated Coastal Zone Management" In: L. G. Luna (ed), *CARICOSTA 2003 - 1st International Conference on Integrated Coastal Zone Management ICZM*, University of Oriente, Santiago du Cuba, May 5-7, 2003.

Khan, M.I., and Islam, M.R., 2003b. "Ecosystem-Based Approaches to Offshore Oil and Gas Operation: An Alternative Environmental Management Technique", Annual Technical Conference and Exhibition, Society of Petroleum Engineer, Denver, USA. October 6-8, 2003. SPE Paper: 84058.

Khan, M.I. and Islam, M.R., 2004. "Assessing Environmental Fate and Behaviour of Oil Discharges in Marine Ecosystem: Using Fugacity Model". In: Proceedings of *the Offshore Oil and Gas Environmental Effects Monitoring Workshop: Approaches and Technologies*, Bedford Institute of Oceanography, May 26 - 30, 2003. Armsworthy, S.L., P.J. Cranford, and K. Lee (Eds.), Battelle Press, 145-165.

Khan, M.I. and Islam, M.R., 2005a, "Environmental Modeling of oil discharges from produced water in the marine environment", Paper no.: GC-225, 33rd Annual General Conference of the Canadian Society for Civil Engineering, Toronto, Ontario, Canada, June 2-4, 2005 / 2-4 Jun 2005, in press.

Khan, M.I. and Islam, M.R,. 2005b. "Assessing Sustainability of Technological Developments: An Alternative Approach of Selecting Indicators in the Case of Offshore Operations". *ASME Congress*, 2005, Orlando, Florida, Nov 5-11, 2005, Paper no.: IMECE2005-82999.

Khan, M.I. and Islam, M.R., 2005c. "Environmental Fate and Behaviour of Oil Discharges in Marine Ecosystem: Using Fugacity Model". *Third International Conference on Energy Research and Development (ICERD-3)*. State of Kuwait, November 21-23, 2005.

Khan, M.I, Zatzman, G. and Islam, M.R., 2005. "New Sustainability Criterion: Development of Single Sustainability Criterion as Applied in Developing

Technologies". Jordan International Chemical Engineering Conference V, Paper No.: JICEC05-BMC-3-12, Amman, Jordan, 12 - 14 September 2005.

Khan, M.I. and Islam, M.R., 2008. "Petroleum Engineering Handbook: Sustainable Operations", Gulf Publishing Company, Austin, Texas: 461 pp.

Khan, M.I. and Islam, M.R., 2007. "True Sustainability in Technology Development and Natural Resources Management". Nova Science Publishers, New York: 381 pp.

Khan, M.I., Chhetri, A.B., and Islam, M.R., 2007. Community-Based Energy Model: A Novel Approach to Developing Sustainable Energy. Energy Sources: Economics, Planning and Policy, Vol. 2, No. 4: 353 - 370.

Lakhal, S.Y., and H'Mida, S., and Islam, M.R., 2005. "Green supply chain parameters for a Canadian petroleum refinery company", International Conference on Computers and Industrial Engineering. Istanbul, Turkey, June 19-22, 2005, Vol. 2: 1273-1280.

Lakhal, S. Y., and H'Mida, S. 2003. A Gap Analysis for Green Supply Chain Benchmarking. In "32th International Conference on Computers and Industrial Engineering", Vol. Vol. 1, pp. 1: 44-49., Ireland, August 11-13th, 2003.

Lakhal, S. Y., and H'Mida, S. (2004). A model for Assessing the Greenness Effort in a Supply Chain. In "35th Decision Sciences Institute Annual Meeting", pp. 5331-5341. Decision Science Institute, Boston, USA.

Lasschuit, W., and Thijssen, N., 2004. Supporting supply chain planning and scheduling decisions in the oil and chemical industry. *Computers and Chemical Engineering*, Vol 28: 863-870.

Lojan, S., (2005). Personal Communication. Senior Petroleum Geologist, Clipper Energy Supply Co., Houston, Texas, USA.

Malik, Q. M. and Islam, M.R., 2000. "CO_2 injection in the Weyburn field of Canada: optimization of enhanced oil recovery and greenhouse gas storage with horizontal wells"., *SPE paper no. 59327*. SPE/DOE Improved Oil Recovery Symposium held in Tulsa, Oklahoma, 3-5 April 2000.

Neiro, Sergio M.S. and Pinto, J.M., 2004. A general modeling framework for the operational planning of petroleum supply chains. *Computers and Chemical Engineering*, Vol 28: 871-896.

OWTG (Offshore Waste Treatment Guidelines), 2002. National Energy Board, Canada-Newfoundland Offshore Petroleum Board, Canada-Nova Scotia Offshore Petroleum Board. August 2002. ISBN 0-921569-40-8.

Pinder, D., 2001. "Offshore oil and gas: global resource knowledge and technological change". *Ocean and Coastal Management,* Vol 44: 579-600.

Powell, T, 2004. "The UK Offshore Sector – A Situation Report", Business Briefing: Exploration and Production: the Oil and Gas Review 2004. Head of Offshore Division, Health and Safety Executive. 1-4 pp.

Sadiq, R., Khan, F.I., and Veitch, B., 2005. "Evaluating offshore technologies for produced water management using GreenPro-I—a risk-based life cycle analysis for green and clean process selection and design", *Computers and Chemical Engineering,* Vol 29: 1023–1039.

Sarkis, J., 2003. "A strategic decision framework for green supply chain management". *Journal of Clcancr production*, Vol 11: 397-409.

Schroeder, D.M., and Love, M.S., 2004. "Ecological and political issues surrounding decommissioning of offshore oil facilities in the Southern California Bight", *Ocean and Coastal Management*, Vol 47: 21–48.

Wenger, L.M., Davis, C.L., Evensen, J.M., Gormly, J.R., and Mankiewicz, P.J., 2004. "Impact of modern deepwater drilling and testing fluids on geochemical evaluations", *Organic Geochemistry*, Vol 35: 1527–1536.

Veil, J.A., 2002. "Drilling Waste Management: past, present and future". Annual Technical Conference and Exhibition, San Antonio, Texas, 29 Septermber-2 October, 2002. *SPE paper no. 77388.*

WCED, 1987. "Our Common Future. World Conference on Environment and Development". Oxford: Oxford University Press; 1987. 400pp.

Wells, P.G., Buttler, J.N., and Hughes, J.S., 1995. "Introduction, Overview, Issues", Peter G. Wells, James N. Butler, and Jane S. Hughes, (Eds), *Exxon Valdez Oil Spill: Fate and Effects in Alaskan Waters, ASTM STP 1219*, American Society for Testing and Materials, Philadelphia, 1995.

Zatzman, G.M., and Islam, M.R., 2005. "Natural Gas Pricing", *In.* Mokhatab, S., Speight, J.G., and Poe, W.A., editors, Handbook of Natural Gas Transmission and Processing. Elsevier Inc.: *in press*.

In: Sustainable Petroleum Operations
Editor: Rafiq Islam
ISBN 978-1-61324-157-8

Chapter 3

A Novel Environmental Risk Management Model for Petroleum Operations

M. Yousuf Mehedi,[1] M. Ibrahim Khan,[2] Chefi Ketata,[1] and M. Rafiqul Islam[1*]

[1]Department of Civil and Resource Engineering, Dalhousie University
1360 Barrington Street, Halifax, Nova Scotia, B3J-2X4, Canada
[2]Institute for Sustainable Energy, Environment, and Economy
ES602-2500 University Dr, NW, University of Calgary, Calgary, AB, Canada
Email: Ibrahim.Khan@ucalgary.ca

Abstract

Contaminated water is an environmental threat due to its toxicity and potential ecological consequences. Oil and gas operations introduce the enormous amount waste stream in the marine environment. The offshore environment of Eastern Canada is not an exception. Managing this huge volume of water in a coastal domain is indeed a challenge. Fisheries are treated as the valued ecosystem component in the coastal zone of Nova Scotia. The paper examines the different management options in regards

---

* Corresponding author: Email: mmehedi@dal.ca, ibrahim.khan@dal.ca, cketata@dal.ca, and rafiqul.islam@dal.ca*.

to valued fisheries components. It addresses the critical issues in managing the ecological risk generated from produced water to the valued ecosystem component with particular reference to fisheries as important component for the people in the coastal zone of Nova Scotia. An integrated risk management model has been proposed that effectively integrates the possible means to reduce the toxicity and volume and other issues posed by produced water from offshore operations. This model can be applied in any offshore hydrocarbon operations to protect vulnerable valued ecosystem.

Keywords: risk management; valued ecosystem components; offshore operations, petroleum.

1. Introduction

The exploration and exploitation of offshore oil and gas resources, from shallow, near-shore water to deeper water areas raise significant environmental concerns as they destroy the environmentally sensitive areas and its valued ecosystem components (VEC). There is formidable evidence that the offshore oil and gas activities expanded rapidly in more ecological vulnerable areas throughout the world (Khan and Islam, 2007). During the extraction of oil and gas and production operations a variety of wastes released to the marine environment that ultimately disrupt the ecosystem integrity and its VECs, such as fisheries. Produced water is considered to be one of the potential sources of environmental damage (Khan and Islam, 2005a). Produced water is a combination of formation water, brine, injection water, process water and other technological waters (CAPP, 2001; UNEP, 2000; NEB et al., 2002; Patin, 1999). The potential environmental concerns therefore arise from variation in quantities with time as well as its complex and considerably variable composition (EandP Forum, 1994).

The Scotian Shelf ecosystem is considered as a productive and complex with varied myriad of issues (Figure 1). Ecological risks rose from all phases of hydrocarbon operations. The most emerging issues of risk includes: drilling and production wastes, resuspension of seafloor sediments, underwater noise, animal disturbance, supply bases, and accidents. Environmentalists have already outlined the coast of Nova Scotia as a component of environmental concern with regard to the development, operation and deconstruction of the site both offshore and onshore areas (Khan and Islam, 2007; 2008). Valued Ecosystem Components (VECs) are the key determinant to measure the magnitude of impacts during construction, operation and abandonment of the site. Fisheries are identified as an important VEC as most of the people in the coastal zone earn their livelihoods through fisheries. The fisheries in the coastal zone of Nova Scotia decline day by day. It is believed that, present oil and gas development project activities are one of the main causes of this degradation. Fisheries have great impact on socio-economic condition in this region. Fishing in Nova Scotia is considered as Canada's oldest economic sector (NHNS, 2002; SCF, 1993).

The fisheries productivity of the Scotian Shelf is considered to be moderately high (DFO, 2002). According to DFO (2002) the Scotian Shelf is treated as Category II ecosystem with 150-300 gC/m^2-yr productivity. Therefore, fisheries can be considered as a VEC. This paper examines the priorities of issues for the management of ecological risks of produced water with special reference to fisheries in Nova Scotia as the valued ecosystem components. It also suggests a new risk management model for valued ecosystem component.

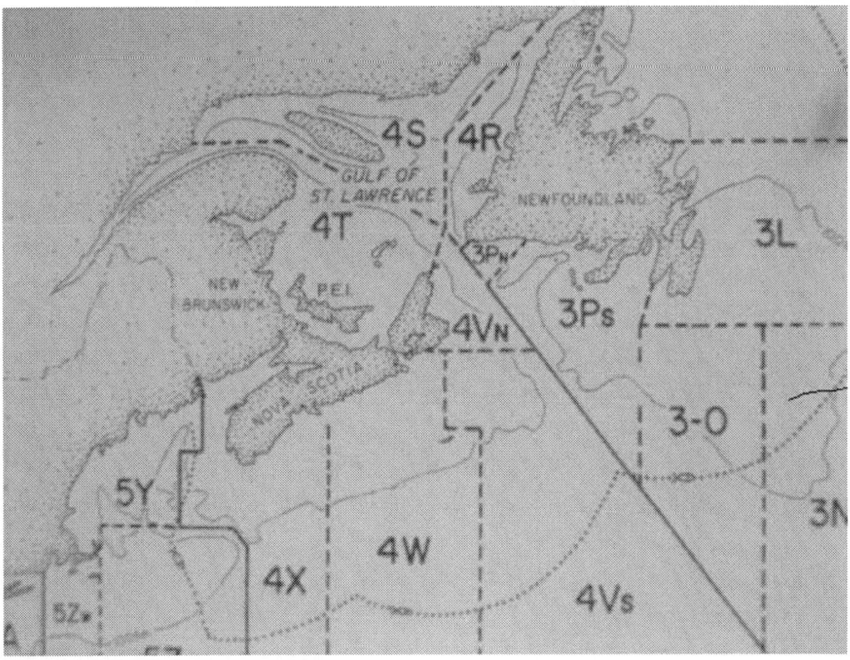

Figure 1. Map of Scotian Shelf (MacKenzie, 2002) www.fisherycrisis.com/images/map.jpg.

2. Risk Management

Risk management is the process of handling a hazardous situation qualitatively and quantitatively. In order to guarantee a project safety and reliability and minimize its losses, it is vital to identify, assess, and eliminate all possible risks in the located area. The objectives of risk management are:

(1) To control and eliminate risks;
(2) To minimize uncertainty regarding the decision-making process;
(3) To ensure the public credibility of the risk management decisions;
(4) To guarantee a safe, reliable, and sustainable environment.

Risk management comprises the following steps (see Figure 2):

(1) Risk identification;
(2) Risk assessment;
(3) Strategy implementation;
(4) Monitoring;
(5) Evaluation.

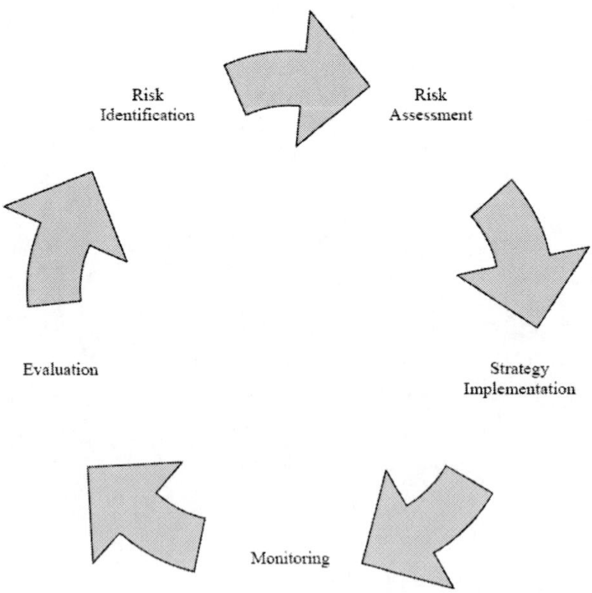

Figure 2. Risk management steps.

3. VALUED ECOSYSTEM COMPONENT MANAGEMENT

As mentioned earlier, fisheries are identified as a VEC. To manage this VEC, there are a number of ways, including restriction on inputs, Environmental Impact Assessment (EIA), Environmental Risk Assessment (ERA), zoning by use, risk-based regulatory approaches, and restrictions on the extent or timing of certain activities (GESAMP, 2001b). The sustainable management of valued ecosystem component requires a series of measures like source reduction, modification of contaminant concentrations, impact reduction and mitigation measures (GESAMP, 2001b; Khan and Islam, 2007). ERA is an important decision making tool that facilitates the management of valued ecosystem components while providing the factual information about the issues of ecological consideration (Landis and Yu, 2000).

In the decision-making stage, the risk assessor provides the risk manager with a full view of the scientific concerns of the different actions, so that they are taken into account during the decision-making process. Therefore, there should be stronger cooperation between risk assessors and risk managers with regards to managing the valued ecosystem components. Figure 3 shows the possible interlink between risk assessor and manger. Risk management process generally integrates information from different areas, such as ecological, engineering, political, economic, and social sciences (Hester and Harrison, 1998). The risk assessment is a part of risk management, and there are different approaches involved in different decision-making contexts. Risk assessment is the scientific aspect of risk management, which should be separated from the social and political aspects of it (Hester and Harrison, 1998).

Risk assessment could be better integrated with risk management by setting the priorities that the risk assessment process could addresses effectively to facilitate decision making. For example, if a risk assessment is conducted on produced water impact on fisheries in Nova Scotia, the process will determine the relative concentration of specific contaminants in valued ecosystem component, i.e., fisheries. However, the risk manager is more interested in subjective factors such as behavior and morphology of the organisms. Risk assessment uses quantitative factors, such as the relative concentrations of contaminants, compared to subjective factors in which the risk manager is interested.

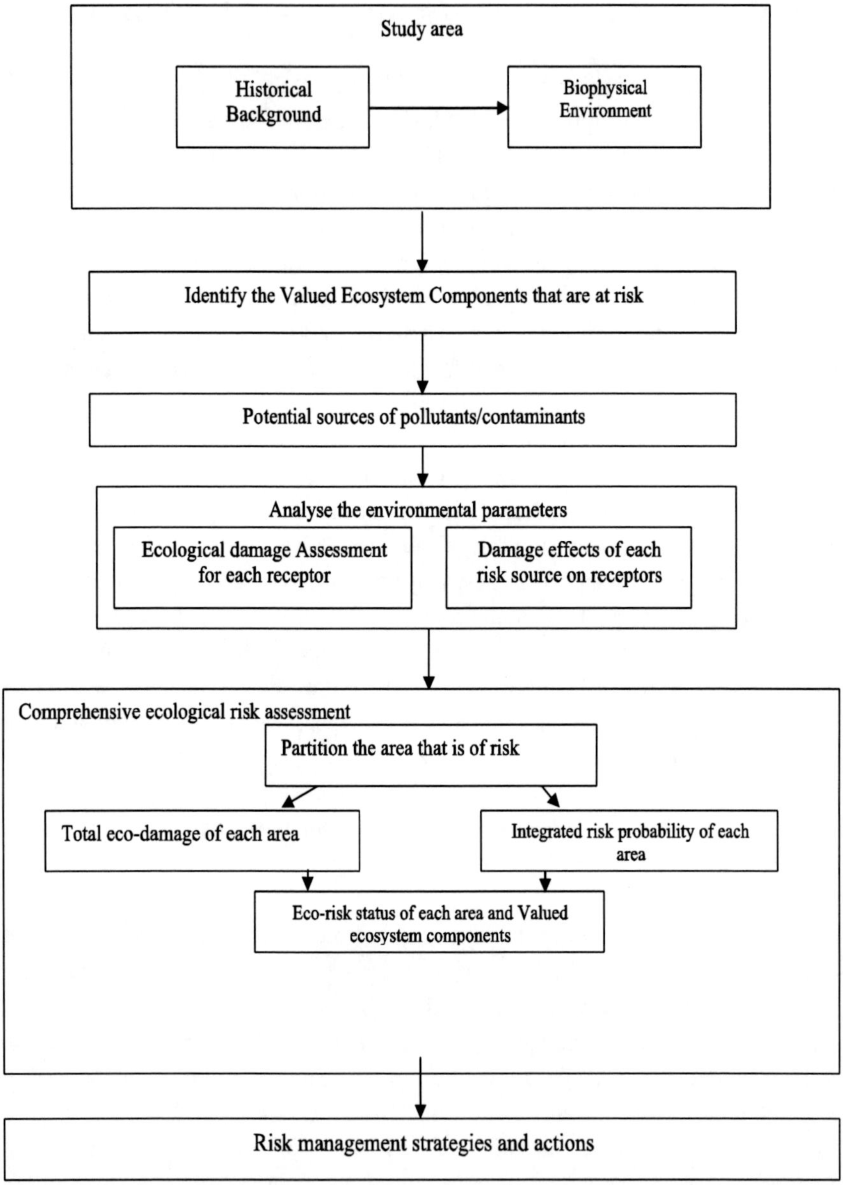

Figure 3. Process of ecological risk management for Valued Ecosystem Component.

4. STRESSOR CHARACTERISTICS

Produced water represents less than 0.4% of the total content of petroleum hydrocarbons entering to the marine environment of the world from all sources (CAPP, 2001; Khan and Islam, 2007). To understand the characteristics and effects of produced water, it is crucial to consider the fate of individual components and how their concentrations vary with the physical and chemical mechanisms. In Figure 4, fate of produced water is shown. The major steps of natural process are identified as dilution, volatilization, chemical reaction, adsorption on suspended solids, and biodegradation (Stephenson, 1992; Khan and Islam, 2005). According to field experiments it appears that produced water dilutes rapidly in the marine environment (Continental Shelf Associates Inc., 1997). However, the impacts of produced water vary depending not only on its quantity and chemical composition, but also on the ability of the environment to absorb, use, or inactivate the constituents of produced water. The impact of dispersed oil in the aquatic environment depends on the location of the discharged point relative to the water column (Stephenson, 1992). If the discharge point is close to the bottom, then local sediment contamination will occur, with associated effects on the benthic community.

Dilution is influenced by the energy of the water, which is expressed as waves, currents, and vertical mixing. When dispersed oil droplets reach the surface, the light aromatic, less toxic compounds volatilize and are lost, with only heavy complex hydrocarbons remaining. It should be mentioned here that the environmental impact of dispersed oil is not very detrimental as most of it is removed with treating equipment generally designed to remove discrete droplets of oil from the water phase. It sometimes increases biological oxygen demand (BOD) particularly in the mixing zone (Stephenson, 1992; CAPP, 2001).

Depending on the chemical characteristics of the component of soluble oil in produced water, its fate varies in the environment. Light, one ringed aromatic hydrocarbons, including; benzene, toluene, xylenes and naphthalines, are characterized by simple, unstable structures that break down or volatilize and dilute quickly (Continental Shelf Associates Inc., 1997). Generally, these compounds do not accumulate in sediments as they are easily broken down (Stephenson, 1992). Similarly, phenols are diluted rapidly, as they are degraded by bacteria and sunlight, both of which convert them into simpler hydrocarbons, which can be further metabolized to produce carbon dioxide and water. However, higher molecular weight aromatic and aliphatic hydrocarbons, may be found in elevated concentrations in sediments near produced water discharge sites, as they are more resistant to degradation, and persist longer in the environment. Dilution is less rapid when a large volume of high saline water is discharged in estuarine, low saline and poorly mixed water. These factors, in combination with relative concentrations of each of the constituents in produced water, contribute to the fact that no adverse impact on the environment has been observed from the soluble oil component of produced water (CAPP, 2001; Stephenson, 1992). When produced water of lower salinity is discharged into the sea the impact would be the same as rain (Stephenson, 1992). Even when water of higher salinity is discharged into the sea, rapid dilution occurs having minimal impact on the local environment.

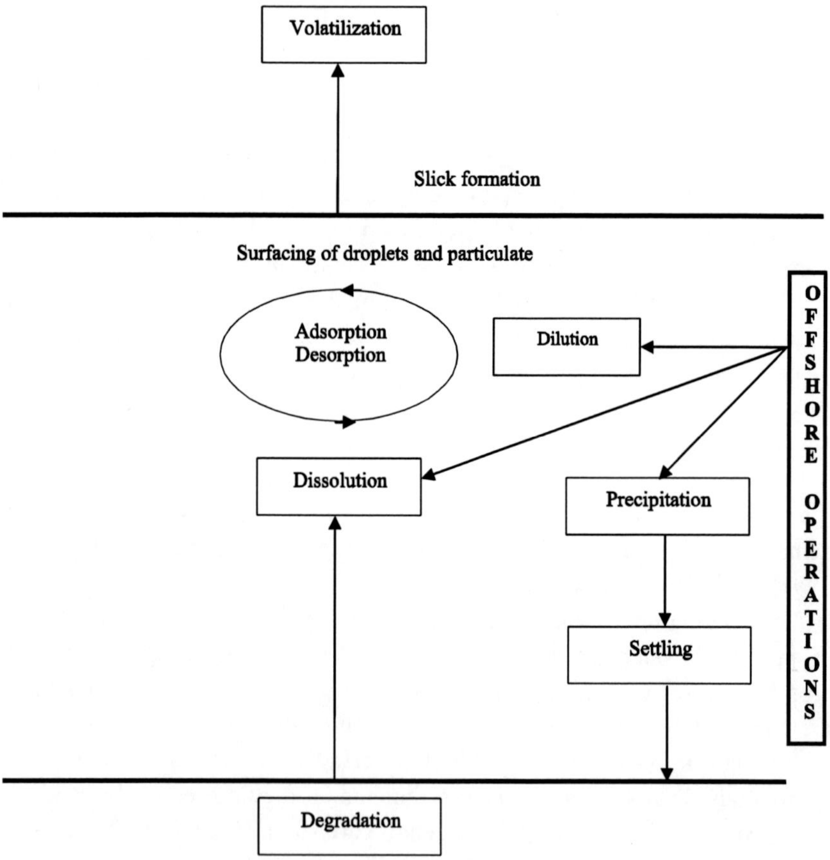

Figure 4. Fate of Produced Water (Modified after Cranford et al., 2001).

Barium from produced water precipitates as barite (barium sulfate) as it mixes with seawater, which is rich in sulfate (Continental Shelf Associates Inc., 1997). Soluble forms of iron and manganese that are present in produced water at high concentrations tend to precipitate as insoluble oxyhydroxides when anoxic produced water mixes with oxygenated natural water. Many other metals present in produced water tend to coprecipitate with barium sulfate, for example. Iron and manganese oxyhydroxides are removed rapidly from the water column similarly. In high energy or deep-water environments it is unlikely that metals from produced water will accumulate in concentrations exceeding background levels, as they are likely to dilute and dissipate rapidly.

Produced water is initially anoxic to prevent operational problems resulting from the presence of oxygen. In high-energy environment, the low oxygen content of produced water does not seem to be a problem. However, in low-energy environments problems associated with anoxic waters are more prevalent. In order to avoid such environmental strain, anoxic produced water should be controlled, and discharged above the surface of the receiving water, enabling some oxygenation to occur prior to entry into the aquatic environment. Thus, while some of the constituents of produced water have lesser impact than others, some can have serious toxic impacts depending upon their concentration, quantity, and method of discharge. The toxic constituents of produced water have to be closely monitored and treated to ensure unacceptable concentrations are not discharged into the marine environment (CAPP, 2001).

5. Risk Evaluation Methodology

5.1. Risk Quotient Model for Estimating the Risks

To measure the ecological risk we followed the deterministic approach by USEPA called quotient method. It is accomplished through comparing toxicity and exposure to the environment. Risk quotient (RQ) is calculated by dividing a point estimate of exposure by a point estimate of effects. This ratio of the calculation is usually used to quantify extent of ecological risk in a certain environment. The formula is as follows:

Risk Quotient (RQ) = Exposure / Toxicity (1)

In determining the risk quotient all the environmental parameter such as environmental fate, transport. Exposure to the ecosystem and the data of ecological effects are considered. The estimated environmental concentration is compared to an effect level in the applied quotient method, for example an $LC_{50,}$ (the concentration of a pollutant where 50% of the organisms die) (Khan and Islam, 2007).

5.2. Analysis of Level of Concerns (LOCs)

Levels of Concern (LOCs) are trigger ratios used by regulatory authorities for comparison against calculated Risk Quotients (RQs). LOCs is the key to measure the ecological risk in the process of ecological risk assessment. LOC effectively address the uncertainties due to possible exposure from human activities and estimated environmental concentrations (USEPA1999). It helps in measuring the status eco habitats by evaluating the concentrations of contaminants of concern. Fisheries as mentioned a valued ecosystem component in the Scotian shelf. With the LOC value we can easily estimate the level of concern and examine the potential environmental consequences due to offshore oil and gas operations Thus, there remains the possibility that untested VEC may be more sensitive to a particular pollutant than those tested. The LOC trigger values established by USEPA is a standard measures to ensure adequate protection for more sensitive, untested, and endangered species (USEPA, 2000, Khan and Islam 2007). LOC's values are used to measure whether VECs are acute or chronic. LOCs are regulatory triggers used to categorize whether the potential risk is of low, medium, or high concern. For instance, a risk quotient less than 0.5 developed for an acute response is of minimal concern to VECs, while a quotient of 0.5 or greater suggests potentially higher acute risk. If the risk quotient is categorized as minimal for a chemical where no LOC is triggered, the use of produced water is predicted to cause no adverse effects for the valued ecosystem fisheries.

Produced water in the coastal zone of Nova Scotia with risk quotients that generally did not exceed the LOC supported by extensive risk mitigation requirements prior to detail risk management. Table 1 presents the levels of concern of VEC, due to release of produced water from offshore operations on the Scotian Shelf. Optimally, mitigation efforts lower potential risk concerns below the LOC, frequently through a refined Environmental Effects Concentrations (EEC). Reducing the LOC by applying different natural adsorbents such as teawaste, cost effective treatment technologies, can be used to minimize the risk to fish exposure as a valued ecosystem component in the coastal zone of Nova Scotia.

Table 1. Level of concern for the valued ecosystem components

Level of Concern (LOC)			
Valued Ecosystem Components (VECs)	Risk Quotient (Exposure/Toxicity)	Values	Risks
Cod	EEC/LC_{50}	0.4	low
Flat fish (e.g. flounder, plaice and halibut)	EEC/LC_{50}	0.5	low
Red fish	EEC/LC_{50}	0.4	low
Haddock	EEC/LC_{50}	0.4	low
Atlantic herring	EEC/LC_{50}	0.5	low
Atlantic Tuna	EEC/LC_{50}	0.5	low
Atlantic mackerel	EEC/LC_{50}	0.3	low
Snow crab	EEC/LC_{50}	0.4	low
Northern shrimp	EEC/LC_{50}	0.3	low
Short fin squid	EEC/LC_{50}	0.3	low

The data and assessment of produced water from offshore operations in Nova Scotia indicated that it does not exceed LOCs of acute and chronic exposures with special reference to fisheries as a valued ecosystem component. Thus, no additional testing or mitigation is required. For more toxic water, the risk quotients are greater than the threshold for presumption of risk for both acute and chronic exposures. Additional testing, or more extensive evaluation of the Environmental Effects Concentrations (EEC), may be required to demonstrate a reduced risk, or risk mitigation measures might be adopted.

6. Issues in Designing a Management Model

6.1. Reduction of Volume

Among the volume of waste released from offshore operations produced water is the highest and the coastal zone of Nova Scotia is not an exception. Indeed production of oil is responsible for generating this huge volume of produced water. Produced water volumes associated with gas production are usually significantly smaller. Figure 5 shows the amount of produced water in Sable Energy Projects on the Scotian Shelf from 2004 to 2005. The differences in the natural mechanisms of production combined with the physical behavior of the fluids are the main factors that reduce the volume in this case. The data provided below show that these patterns appear to be true for Canadian fields as well (CAPP, 2001). The composition of produced waters from Canadian offshore fields is complex, as it is in other parts of the world. In fact, based on available data there appears to be little of a unique nature associated with Canadian produced waters. Results from studies performed in other parts of the world can be applied to offshore Canada, when considered in the proper context.

Oil and gas resources of offshore Canada are in the early stages of development with only three discoveries having been produced to date. Table 2 summarizes the production histories of the fields to date. Produced water discharges have occurred at each of the fields. Discharges at Cohasset ceased with production in December 1999. Discharges from Hibernia and Sable Island continue today. Effluent monitoring is occurring as stipulated by the Offshore Waste Treatment Guidelines, and produced water treatment systems are performing as expected.

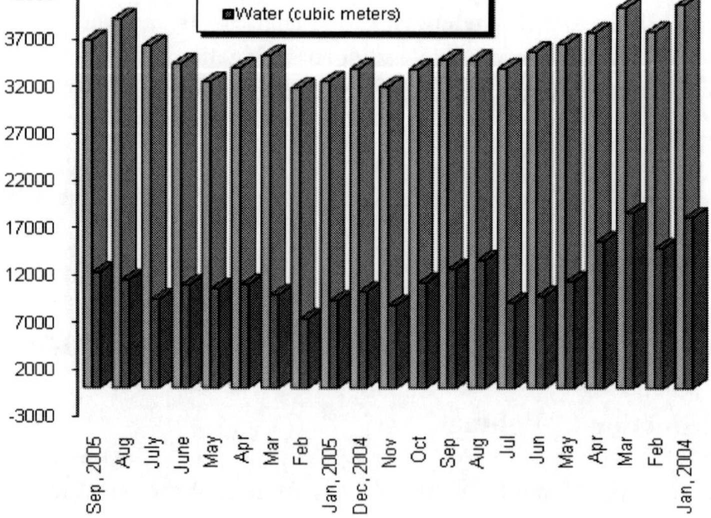

Figure 5. Average Monthly Production of raw gas (ten thousand cubic meters) and produced water (cubic meters) in the Sable Energy Project from January 2004 to September 2005 (Khan and Islam, 2007).

Table 2. Number of oil-producing installations in UK and quantities of produced water discharged from these installations (adapted from UKOOA, 1999)

Year	Number of oil-producing installations	Produced water quantity (tonnes)
1996	59	210,000,000
1997	64	234,000,000
1998	67	253,000,000
1999	67	260,000,000

It is important to measure the impacts of the constituents of produced water during the risk assessment process. According to UNEP (2000) volume of produced water depends on factors such as, permeability of the reservoir rock, underground pressure, viscosity of the hydrocarbon, and oil/gas ratio. There is alarming evidence that the release of this high volume of produced water accompanied with series of contaminants could impede and/or disrupt the fisheries habitat.

To visualize the magnitude of quantities of produced water involved in offshore drilling operations, the case of United Kingdom (UK) is examined. Tables 2 and 3 show the quantities of produced water released from UK offshore oil and gas installations over four years. These tables show that there is a steady increase in the production of produced water. Furthermore, they show that gas reservoirs generate much lower amounts of produced water compared with that of oil reservoirs. In designing effective management plan the environmental policy maker should choose and apply alternative and cost effective technologies in the following stages:

- Determination of the type of reservoir, i.e., oil or gas.
- Determination of the type of rock.
- Determination of the size of the reservoir.
- Determination of the stage of production.

The environmental effects on Valued Ecosystem Component along the temporal and spatial scales rely on the volume of produced water. The guideline for produced water discharges often does not work as produced water dilutes rapidly. If the manager could measure the amount of produced water expected to dilute in that area, then it would be simpler to design an effective management strategy considering open and closed water masses, and existing or nonexisting mechanisms of the different chemical processes.

Table 3. Number of gas-producing installations in UK and quantities of produced water discharged from these installations (adapted from UKOOA, 1999)

Year	Number of oil-producing installations	Produced water quantity (tonnes)
1996	3	618,000
1997	24	328,000
1998	32	451,000
1999	33	1,000,000

6.2. Reduction of Toxicity

Moffitt et al. (1992) collected twenty four produced water samples from four "low" discharge rate (<10,000 bbl/day) platforms and from four "high" discharge rate (>10,000 bbl/day) platforms in the Gulf of Mexico. Seven-day toxicity tests utilizing Mysidopsis bahia (mysid shrimp) and Cyprinodon variegatus (sheepshead minnow) were conducted on each sample using procedures specified by the US EPA for short-term chronic toxicity tests using marine species (CAPP, 2001). To assess potential temporal variability in toxicity, three samples were collected from each platform and sampling times were separated by approximately two weeks. This program was designed specifically to evaluate toxicity testing requirements that were being enacted by EPA for the gulf of Mexico. In addition to the 96-hr LC_{50} and the 7-day LC_{50}, No Observable Effect Concentrations (NOECs) were determined for survival and growth in both test species and for fecundity in the mysid shrimp in seven-day tests.

To determine if toxicity differed significantly as a function of sampling time or discharge rate, statistical comparisons were made using Two-Factor Analysis of Variance (ANOVA) procedures and Tukey's method of multiple comparisons. All statistical comparisons were conducted at $p<.05$ level of confidence. All toxicity endpoints were included in the analyses. No significant differences were observed between results from sample collected at different time periods or from offshore platforms with varied discharge rates or between any combinations. The results are summarized in Table 4.

Table 4. Examples of effects of produced water, studied under field conditions (Adapted from Wills, 2000)

Concentration	Test Duration	Test Organisms	Effects	References
Dilution of 25μg/l	Hours and days	Larvae of Atlantic cod (North Sea)	Reduced Survival	Davies *et al.*, 1981
0.15% with 5-15 μg/l of oil	100 days	"Natural zooplankton" (North Sea)	Reduced survival at population level	Davies *et al.*, 1981
5%	48 hours	Larvae/embryos of oyster Crassostrea gigas (North Sea)	50% mortality	Somerville *et al.* 1987)
5-6%	15 minutes	Luminescent bacteria (Noth Sea)	50% decrease in photoluminescence intensity	Somerville *et al.* 1987)
5-6%	48 hours	Daphnia sp. Daphnia magna	Immobilization of 50% of test organisms	Somerville *et al.* 1987)
10%	24 hours	Trout Salmo gairdneri	50% mortality	Somerville *et al.* 1987)
0.1%-0.3%	60-130 days	Natural communities of plankton in mesocosms (North Sea)	Reduced abundance of copepod larvae, disturbances of trophic structure, biochemical changes in fish larvae	Gable *et al.*, 1987
3-30%	96 hours	Invertebrates and fish (Gulf of Mexico)	50% mortality	Middleditc h 1981
1-10%	Chronic tests	Macrophytes Macrocystis pyrifera (California Shelf)	Reduced ability of zoospores to settle on the bottom	Lewis &Reed 1994
20-70mg/l	1-2 days	Fish eggs, fry and larvae (Caspian Sea)	Death of juveniles at 60-70 mg/l; larvae at 35-40mg/l; eggs at 20-25mg/l	Alkperov 1989
2-30 mg/l	"Several days"	Invertibrates (Caspian Sea)	Death of crustaceans at 30 mg/l; reduced survival at 2 mg/l	Alkperov 1989

Endpoints for the mysids ranged from 7.08% for the 96-hr LC_{50} to 1.6% for the NOEC growth rate. Endpoints for the sheepshead minnows ranged from 21.5% for the 96-hr LC_{50} to 2.75% for the NOEC growth rate.Also shown in Table 4 are values from the Louisiana Department of Environmental Quality's database of approximately 400 produced waters tested in its program. The data from the two studies are in general agreement (CAPP, 2001).

Reducing the toxicity from produced water should be high priority for a risk manager. It is important to know how the mechanism of produced water affects the valued ecosystem component. In designing appropriate measures for toxicity the manager needs appropriate toxicity data from the assessment of valued ecosystem component.

There are three basic ways to assess the toxicity of mixtures:

1. If the mixture has a reasonably stable composition and is available to the assessor, the most realistic approach is to assess the toxicity of the whole mixture. If the results elevated toxicity, further analysis is needed. Each of the mixture components is tested and those that occur in toxic concentrations are removed. If no such component is revealed, the mixture is fractionated and the fractions are tested, analyzed, and possibly refractionated until the toxic fractions are identified and treated (Suter, 1993).
2. If the initial mixture has relatively few components or the toxic effects are known to be due to few components in varying concentrations, then synthetic mixtures are tested. The advantage of this method is that the combined impact can be isolated and used to predict the effects of variability in the mixture. Synthetic mixtures are used to confirm the results of fraction mixtures, used in the previous approach (Suter, 1993).
3. If the assessor cannot obtain test data for a mixture, he makes extrapolations from existing data. Additivity is assumed to occur among the components of the mixture. This approach is justified by studies of acute lethality. However, the assumption of additivity impact of a mixture is not always realistic as it depends on the duration of exposure and on the response being measured (Suter, 1993). In addition, it is often unrealistic to compare results of different studies because of the use of different reference models with

inherently different operational definitions of additivity, synergism, and antagonism (Sharma *et al.,* 1999).

6.3. Monitoring

By various forms of monitoring, scientists can provide environmental managers with information on whether individual operators are complying with regulatory requirements and whether the condition of the marine environment is improving or deteriorating. The design of monitoring programmes, therefore, must not simply involve, but be driven by, managers rather than scientists (GESAMP, 2001a).

In the case of produced water, monitoring should be based on certain measures that could identify environmental impacts on valued ecosystem components. This should include sampling of eggs, larvae, juvenile fish, sediments, and water. In addition, measurements of wind and currents can give the exact location and extent of environmental damage. Adult commercial species should also undergo laboratory and taste-panel tests to determine whether fish are tainted. If fish are tainted or if there is a loss to the fishery because of the perceived tainting of fish, then compensation may be paid to the affected fishers.

Considering the significant environmental effects of produced water (see Table 4), an effective monitoring program is needed to facilitate the risk management process. A site and organism-specific monitoring program will make the full assessment of produced water possible. There are management reasons embedded in the ecological risk assessment that justify a site and organism-specific monitoring program.

A site and organism-specific monitoring program will enable the risk manager to choose how he will reach conclusions about the impacts on valued ecosystem components. There are two general approaches in monitoring (Peterson et al., 2001): a data-based approach, where new data is acquired and the risk manager directly tests the impacts, and a model-based approach, which includes no collection of new data, but involves the deduction of a conclusion from a model.

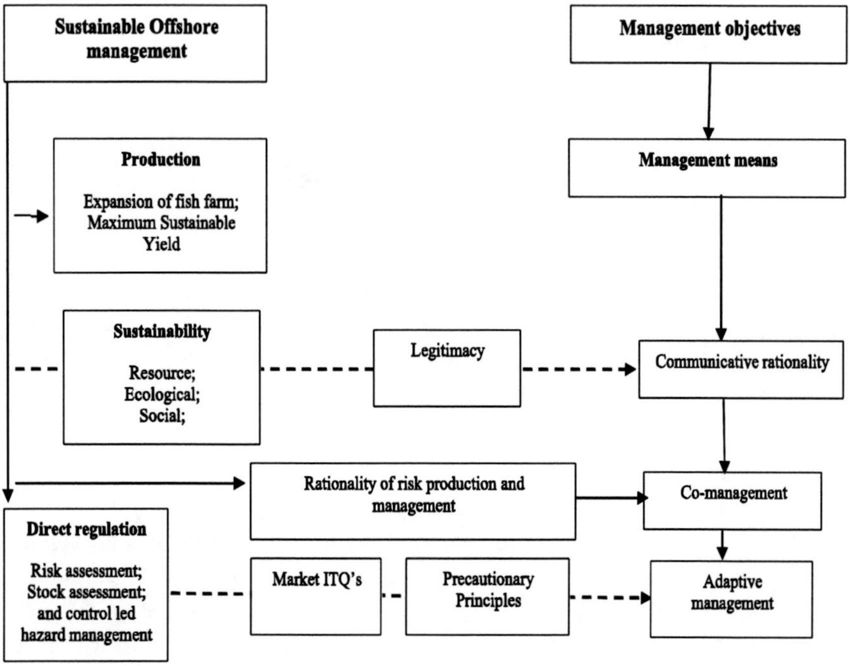

Figure 6. Proposed risk management model for sustainable offshore operations.

The monitoring program has the design considering the results of risk assessment and based on the decision from risk manager. Site and organism-specific monitoring has been conducted to measure the threatening impacts from oil spills (Dyrynda et al., 2000) and contamination in a specific marine area. It is obvious that the more specific the monitoring, the more likely it is to reveal complex effects that are time, site, and, organism-influenced. Another advantage of this type of monitoring program is that it provides the risk manager with more valid data on the organisms of interest.

In designing an appropriate monitoring program, the manager needs to prioritize in selecting appropriate endpoints and should consider the fate and effects of produced water. The ecotoxicological test systems that are used include: (1) laboratory short-term tests for acute toxicity, with mortality of the organism as the test endpoint (EandP Forum, 1994). These tests have greater value when assessing effects on organisms that are exposed to rapidly decreasing concentrations of produced water, like planktonic organisms, and organisms that can avoid contamination by moving away, such as fish and shrimps; (2) longer term laboratory tests for chronic toxicity, with sublethal effects on the organisms that are exposed to contaminants for extended periods, such as benthic organisms living close to discharge points; and (3) field mesocosm experiments that are used to assess possible effects at the community level and are important to identify indirect effects. Tests are conducted on a range of aquatic organisms of the basic three trophic levels to assess the effects of produced water in marine environment. These tests include both native and nonnative species of the study area, species that are well accepted internationally as useful indicator species in ecotoxicology.

The monitoring should be designed and conducted as accurately as possible with special reference to the data collection process to facilitate ecological risk assessment. The most effective monitoring to determine the effects of produced water on aquatic organisms during offshore drilling would be an organism and site-specific monitoring program that takes into account the site characteristics of the area and the organisms that live there. Even though a monitoring program might seem a straightforward process, its actual design is quite complex. The considerations that the risk assessor and the risk manager have to incorporate when designing a site and organism-specific monitoring program are summarized as:

1. Spatiotemporal design.
2. Total area sampled.
3. Boundaries of the area of environmental concern.
4. Treatment of habitat heterogeneity within sites.
5. Selection of appropriate organisms based on scientific criteria.
6. Philosophical support for choosing certain organisms (i.e., importance for the area).
7. Scope of the assessment for the communities and habitats.

Risk assessor and risk manager should be careful as regards to the points of sampling and the spatiotemporal frame, sampling boundaries to make the monitoring effective. The risk manager is constricted by political boundaries that seldom match the ecological boundaries. Site-specific monitoring program should consider both political and ecological imperatives in designing the process. The sampling and the monitoring in general have to be designed in a way that it incorporates the habitat heterogeneity within sites. The above reasons make the selection of the site of samplings and the selection of appropriate organisms a thoughtful process, rather than a coincidental one based on the data that already exists. The selection of appropriate organisms should be supported by a theory. The basic consideration of a risk manager's choice is the importance of certain organisms while that of the risk assessor is ensuring that organisms are appropriate ecosystem indicators and cover different trophic levels in the marine environment. The matching of these choices can result in an effective organism-specific monitoring mechanism. It is important to incorporate the scope of the assessment for the human and nonhuman communities of the area when designing a site and organism-specific monitoring program.

6.4. Legal Analysis

International Commission for the Northwest Atlantic Fisheries (ICNAF) regulates the fisheries of Scotian Shelf since 1945. On January 1, 1977, following a series of lengthy international negotiations at U.N. Law of the Sea Conference, Canada's jurisdiction over coastal waters was extended to 200 nautical miles. Since the extension of fisheries jurisdiction, the Department of Fisheries and Oceans (DFO) has regulated the fisheries of Nova Scotia (SCF, 1993)). The Ministry of Fisheries introduced a series of management plans, to conserve the fisheries, such as Community-Based Management, ITQ Fleets, and popular TAQ (DFO, 1999). These management plans provide commitment to community preservation.

The 1982 Law of the Sea Convention requires that catch possibilities in addition to domestic needs be made available to foreign parties. This gives the commitment, not only to Canadians, but also to other nations to access the fisheries. For example, in recent years, foreign (Cuban and Russian) participation in the Scotian shelf silver hake fishery has been allowed, as required under this provision (DFO, 2002). According to treaty rights and constitutional commitments, an Aboriginal Fishery is also active.

The Canada-Nova Scotia Offshore Petroleum Board (CNSOPB) is responsible for protection of the environment during all phases of offshore petroleum activities, from initial exploration to abandonment (CNSOPB, 2002). Under the Canadian Environmental Assessment Act, the CNSOPB is now officially a federal authority. It is also involved in initiatives led by the DFO related to marine protected areas and integrated management planning under the Oceans Act (CNSOPB, 2002). While some other federal departments deal with fisheries matters, the DFO takes chief responsibility of water and its resources.

6.5. Spatial and Temporal Impact Assessment

Although offshore oil and gas production operations using confined platform drilling rigs are located within the shallow coastal zone of Nova Scotia, the impact of produced water could be observed far away from the coast. The migratory behavior of the pelagic fish is the main reason for the transfer of pollutant to a greater distance, as many fishes spend some of their life cycle on the Scotian Shelf. The effects of produced water and pollution may extend further than the legal boundary. If there is constant release of effluents from produced water, phytoplankton and zooplankton production will be destroyed, which will severely affect the plankton feeding fish and fish larvae through food chain. The effects on the productive coastal and estuarine system will be harmful for the overall marine ecosystem.

The interaction of oil and gas development projects with fisheries has large scale effects. Effects might extend for the short term exploration or construction activity during breeding and migration periods of some fish species. Herring, tuna, and mackerel are the most migratory fishes. Mackerel approach the Atlantic coast in late May in large schools and leave again in the fall for over-wintering (NHNS, 2002). Bluefin tuna spend winters in warm southerly waters and move to Nova Scotia as the season progresses. If the seismic operation or other exploratory operations are taking place in this period then the effects will be greater. The overall effects on fisheries will be continuous throughout the lifecycle of an offshore oil and gas project.

7. Proposed Risk Management Model

Considering the development of the basic objectives and means of risk management strategies, a new management model is developed, which is shown in Figure 7. Details risk management objectives and proposed steps are schematically outlined in this figure. This model takes into account the broad social context along with the valued ecosystem fisheries.

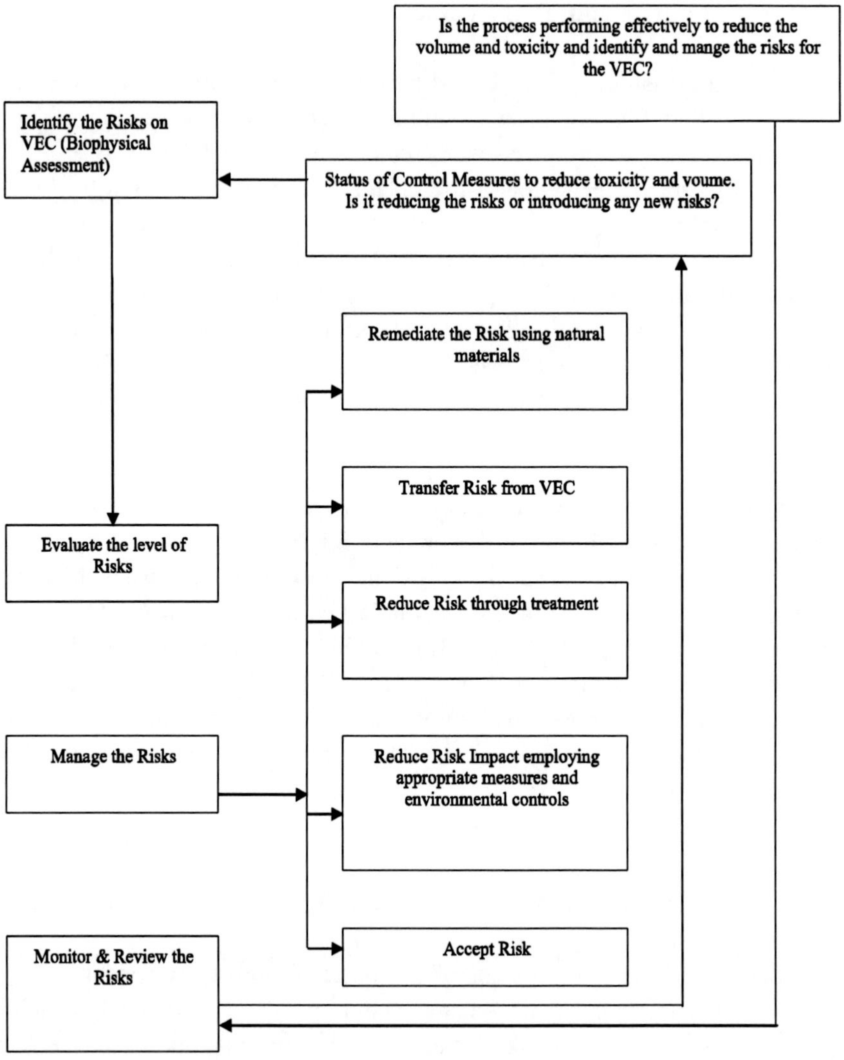

Figure 7. Risk Management Model.

The environmental objectives are clearly included in the model to achieve sustainability with regards to VEC. The principle rationale comprises to increase fisheries production by implementing fish farming and maintaining maximum sustainable yield. This measure ensure sustainability through environmental control measures as well as through input (effort) and output (Total Allowable Catch) control within the realm of sustainable offshore management. The model is designed by establishing links between environmental indicators and sustainable offshore development. In the model, legitimacy is also considered as an important component. In addition, various comanagement schemes have been suggested as the appropriate means of achieving sustainable risk management of valued ecosystem fisheries in Nova Scotia.

As fisheries are important component of livelihood in coastal Nova Scotia. Therefore, social value is an important priority in managing ecological risk posed by the offshore production operations that are neglected in the existing management models. Generally adaptive management process includes various political, social, cultural and biological domains, but the basic approach is in any case that the reactions of certain components (social, biological, or both) to specific management means are unpredictable and management must adjust to the response. Such schemes may include aspects of both risk management and increased legitimacy. The proposed risk model (see Figure 7) for the VEC focuses on the objective and means in relation to sustainable management.

There is always a gap in the information available in designing a management model, and technical deficiencies in the available tools. Gaps in understanding the consequences of management actions and the lack of tested techniques for integrating multiple objectives and risk in decision making hamper efforts with these new challenges. Natural adsorbents can be used to remediate the toxicity of produced water (Al-Darbi et al., 2002). Transferring risk from valued ecosystem component and employing environmental regulations and cost effective treatment is another way to mitigate risk from offshore operations.

As the main objective of this paper is to develop an integrated model, test the integrated ecological assessment and decision methodologies, as well as monitor and review the severity of any risks to protect the VEC that are at risk due to offshore production operations. The proposed model (Figure 8) assists managers and stakeholders who are involved in designing effective risk management plans to know the status of control measures and how to integrate effectively prior issues like volume, toxicity and other concerning issues. The management model is prepared on existing initiatives and addresses key issues and management challenges.

Table 5. Priorities of Issues of Action (2 = Medium priority; 1 = High Priority)

Issues	Priorities	Actions
Volume	1	Strategic environmental protection plan; Additional control measures; Receiving facilities of oil and oily substances at the vicinity of production operations; Increased public understanding and awareness of the problems with regard to the volume of produced water; Stringent regulations; More appropriate technology considering economic technical feasibility
Toxicity	1	Chemical characterization and fractionate the constituents; Applying Ecological Risk Assessment tools; Integrated Canadian ocean policy framework and support mechanisms; Implement marine protected areas in the management plans; Stringent regulations;
Legal Mechanisms	1	Periodical Evaluation; Needs to revise the legislations to protect the VECs considering the current situations; Includes the concepts of "precautionary approach"; Introducing zoning system and marine protected areas;
Stakeholder situation	2	Cooperation among all stakeholders including CNSOPB, Department of Fisheries and Ocean, fishermen, Resident, Industry people; Needs to develop and strengthen an appropriate institutional framework to prevent degradation; Incorporate careful definition of the authority, responsibilities and accountability of specific sectors in to the law;
Monitoring	2	Toxicity evaluation; Accurate data to facilitate ERA; Organism and site-specific; Selection of appropriate organisms;

CONCLUSION

This study examined different components in managing ecological risk of produced water in offshore operations. In this paper, an alternative risk management model was followed, which is based on valued ecosystem component. Fisheries were found to be as an important VEC for the Scotian Shelf. It was observed that oil and gas operations directly and indirectly affect the local fisheries due to the consequences of huge volume of produced water. To overcome the fisheries problem some innovative fisheries management strategies were suggested which are based on the principles of zero-waste offshore oil and gas operations. Some of the guidelines are artificial recruitment to compensate the stock depletion, oil companies should establish their own hatchery to produce fish larvae for stocking in this regard. Since fish breeding grounds are very sensitive, oil and gas operations should not be allowed in and around them. Exploration activities should be completely stopped in the fish migration or spawning season. Introducing marine protected areas around the oil and gas development areas should be encouraged to protect the valued ecosystem components.

The risk management issues are identified as to facilitate and improve the environmental management of produced water. There is a need to determine the approximate amount of produced water that will be released into the environment so that we can get the real view of impacts on the valued ecosystem component. Produced water is a complex mixture and to measure the toxicity of this huge volume of produced water is an important concern in framing an alternative management model. A detailed toxicity assessment process is recommended that includes determination of major contributors to the toxicity of produced water. These factors will ultimately determine the priorities for control and an effective plan for managing the ecological risks imposed by produced water. A site and organism-specific monitoring program for each well is suggested. Managing produced water in the marine domain is a challenge, but by applying proposed ecological risk management process will assist in mitigating the risks posed by produced water.

ACKNOWLEDGMENT

The authors would like to thank the Atlantic Canada Opportunities Agency (ACOA) for funding this project under the Atlantic Innovation Fund (AIF).

REFERENCES

Al-Darbi, M., Muntasser, Z., Tango, M. and Islam, M.R. (2002) 'Control of Microbial Corrosion Using Coatings and Natural Additives', *Energy Sources*, Vol. 24, No. 11, pp. 1009-1018.

Canada Nova Scotia Offshore Petroleum Board (2002) *White Page*, Environmental Protection Board, http://www.cnsopb.ns.ca/Environment/environment.html (Cited: April 21, 2002).

CAPP (2001) *Produced water waste management*, Canadian Association of Petroleum Producers, Technical Report, August.

Continental Shelf Associates Inc. (1997) *Metals and organic chemicals associated with oil and gas well produced water: Bioaccumulation, fates and effects in the marine environment*, Gulf of Mexico Produced Water Bioaccumulation Study.

Department of Fisheries and Oceans (DFO) (2002) *GroundFish. integrated fisheries management plan*, Scotia-Fundy Fisheries, Maritime Region, April 1, 2000-March 31, 2002, Department of Fisheries and Oceans, Moncton, New Brunswick.

Department of Fisheries and Oceans (DFO) (1999) *Canadian atlantic integrated fisheries management plan Bigeye, Yellowfin (Thunnus obesus), (Thunnus albacares), Albacore Tunas (Thunnus alalunga), 1998-1999*, Department of Fisheries and Oceans, http://www.mardfo-mpo.gc.ca/fisheries/res/imp/98othon.html (Cited: April 22, 2002)

Dyrynda, E.A., Law, R.J., Dyrynda, P.E.J., Kelly, C.A., Pipe, R.K. and Ratcliffe, N.A. 'Changes in Immune Parameters of Natural Mussel *Mytilus edulis* Population Following a Major Oil Spill ('Sea Empress', Wales, UK)', *Marine Ecology Progress Series*, 206: 155-170.

EandP Forum (1994) *North Sea produced water: Fate and effects in the marine environment*, Report No. 2.62/204, May.

GESAMP (IMO/FAO/UNESCO/WMO/WHO/IAEA/UN/UNEP: Joint Group of Experts on the Scientific Aspects of Marine Pollution) (2001a) *A sea of troubles: Issues in focus*, GESAMP Reports and Studies No. 70, UNEP, The Hague, Geneva, Nairobi, pp. 1-15.

GESAMP (IMO/FAO/UNESCO/WMO/WHO/IAEA/UN/UNEP: Joint Group of Experts on the Scientific Aspects of Marine Pollution) (2001b) *Land-based sources and activities affecting the quality and uses of the marine, coastal and associated freshwater environment*, GESAMP Reports and Studies No. 71, UNEP, The Hague, pp. 75-126.

Hester, R.E. and Harrison, R.M., eds. (1998) *Risk Assessment and Risk Management*, Cambridge: The Royal Society of Chemistry.

Human Resources Development Canada (HDRC) (2001) *Industry Profile: Human Resources Profile: Fish Products Industries and Services Incidental to Fishing.*

Khan, M.I, and Islam, M.R. (2005) 'Environmental Modeling of Oil Discharges from Produced Water in the Marine Environment', *Canadian Society of Civil Engineers 2005*, Toronto, Canada, June 2-4.

Khan, M.I. and Islam, M.R. (2007) *True Sustainability in Technological Development and Natural Resources Management*, Nova Science Publishers, New York, USA: 381.

Khan, M.I. and Islam, M.R., 2008. *Petroleum Engineering Handbook: Sustainable Operations*, Gulf Publishing Company, Austin, Texas: 461 pp.

Landis, W.G. and Yu, M.H. (2000) *Introduction to Environmental Toxicology: Impacts of Chemicals upon Ecological Systems*, 2nd ed., Florida: CRC Press LLC.

MacKenzie, D. (2002) *The Downturn of the Atlantic Cod (Gadus morhua) in Eastern Canada. What is happening to these fish, and why?*, www.fisherycrisis.com
/images/map.jpg

Moffitt, C.M., Rhea, M.R., Dorn, P.B., Hall., J.F., Bruney, J.M. and Evans, S.H. (1992) *Short-term chronic toxicity of produced water and its variability as a function of sample time and discharge rate*, pp. 235-244.

Natural History of Nova Scotia (NHNS) (2002) *Marine fishes*, Natural History of Nova Scotia, Volume I: Topics.

NEB, CNOPB and CNSOPB (National Energy Board, Canada-Newfoundland Offshore Petroleum Board and Canada-Nova Scotia Offshore Petroleum Board) *Offshore Waste Treatment Guidelines*, Canada, Ottawa, 1996.

Patin, S. (1999) Environmental impact of the offshore oil and gas exploration and production: Waste discharges during the offshore oil and gas activity, http://www.offshore-environment.com/discharges.html, Accessed on 27 May 2001.

Peterson, C.H., McDonald, L.L., Green, R.H. and Erickson, W.P. (2001) 'Sampling Design Begets Conclusions: The Statistical Basis for Detection of Injury to and Recovery of Shoreline Communities after 'Exxon Valdez' Oil Spill', *Marine Ecology Progress Series*, 210: 255-283.

Senate Committee on Fisheries (SCF) (1993) *The Atlantic Commercial Inshore Fishery,* Committee Report, June, http://www.parl.gc.ca/36/1/parlbus/commbus/senate/com-e/fish-e /past_rep-e /93repen1.htm (Cited: April 17, 2002).

Sharma, S.S., Schat, H., Vooijis, R. and Van, H.M. (1999) 'Combination toxicology of copper, zinc, and cadmium in binary mixtures: concentration-dependent antagonistic, nonadditive, and Synergistic Effects on Root Growth in *Silene Vulgaris. Environmental Toxicology and Chemistry*, 18(2): 348-355.

Stephens, S.M., Frankling, S.C. and Anne J. (2000) 'Sublethal Effects of Exposure of Juvenile Turbot to Oil Produced Water', *Marine Pollution Bulletin*, 40(11):928-937.

Stephenson, M.T. (1992) 'A Survey of Produced Water Studies', In *Produced Water: Technological/Environmental Issues and Solutions*, Ray, James P. and engelhardt, F. Rainer, eds., Volume 46, New York: Plenum Press.

Suter, G.W. (1993) *II. Ecological Risk assessment*, Chelsea: Lewis Publishers.

United Kingdom Offshore Operators Association (UKOOA) (1999) *Environment: 1999 Environmental Report*, http://www.oiland gas.org.uk/issues/1999report/enviro99_water.htm, Accessed on 12 July 2001.

United Nations Environment Program (UNEP) (2001) *Offshore Oil and Gas Environment Forum: Background Notes on the Offshore Industry and its Environmental Impacts*, http:///www.natural-resources.org/offshore/ background/bgnote.htm, Accessed on 14 May 2001.

USEPA. 2000. Methods for measuring the toxicity and bioaccumulation of sediment- associated contaminants with freshwater invertebrates. EPA 600/R-99/064, March 2000 2nd Edition.

Warne, M.J. and Hawker, D.W. (1995) 'The Number of Components in a Mixture Determines Whether Synergistic and Antagonistic or Additive Toxicity Predominate: The Funnel Hypothesis', *Ecotoxicology and Enironmental Safety*, 31: 23-28.

Willis, J. (2001) *Muddied Water: a survey of offshore oilfield Drilling Waters and Disposal Techniques to Reduce Ecological Impact of Sea Dumping*, http://www.offshore-environment.com/ospar.html, Accessed on 1 July, 2001.

In: Sustainable Petroleum Operations
Editor: Rafiq Islam

ISBN 978-1-61324-157-8

Chapter 4

AN APPROACH FOR CONFLICT RESOLUTION IN OIL AND GAS OPERATIONS

M. Yousuf Mehedi, Arjun B. Chhetri, Chefi Ketata, and M. Rafiqul Islam*

Department of Civil and Resource Engineering
Dalhousie University
1360 Barrington Street, Halifax, Nova Scotia, B3J-2X4, Canada

ABSTRACT

A persuasive analysis of a complex conflict was carried out regarding oil and gas activities in Cape Breton, Nova Scotia. The introduction sets this particular problem based on literature study, with a number of relevant variable sets discussed. The various stakeholders and their position in oil and gas operations created a number of conflicts. The position of the province is a fair rendition of the view of socio-economic development. Fisheries viewpoint was found very significant. The environmental position with a sharp elaboration of the alliance between environmental and fisheries interests, the probable divisions within the general public were discussed. Analysis and recommendations pull together some of the major axes of conflict and pointed out some of the particular difficulties to be dealt with. Some models have been proposed to resolute the conflict. The excessively sectoral approach of many

---

* Corresponding author: rafiqul.islam@dal.ca.

agencies, even presumably connected ones, envisaged Integrated Coastal Management (ICM) principles as an approach for resolution.

Keywords: conflict; resolution; oil and gas operations.

1. INTRODUCTION

Marine environment comprises oceans, seas and adjacent coastal areas, is an essential feature that supports the life system and plays a significant role for sustainable development. The interaction of numerous resource users in coastal areas become complex in a relatively small geographic context. The potential for conflict in the marine environment is increased in areas of limited coastal space, high population densities, diverse marine and terrestrial habitats in close proximity, and differing economic and social interests. Most conflicts arise from competition for limited coastal space, use of limited coastal resources, inadequate interpretation of information, or from conflicting environmental and economic interests (Suman, 2001). De Bono (1985) describes conflict as an expected and revered idiom of our civilisation that results in change. However, it is most commonly described as a struggle of interests, values, actions or directions, often resulting from individual or group fears (Bercovitch 1984, de Bono 1985, Fisher et. al. 1992, Johnson and Duinker 1993).

Coastal conflicts are commonly multidimensional and may be based on economic, social, legal, institutional and resources interests. Overlapping interests, incompatible uses and differing values (e.g. development vs. nature conservation) can result in conflict between and among authorities (government), resource users and non-consumptive resource users (e.g. environmental groups). The root cause of conflict in coastal zones is often difficult to pin-point.

Conflicts are important, as they are part of the dynamic nature of coastal management and should not always be interpreted as negative. In most instances, conflict can be viewed as an impetus to change (Brown, 1997). Conflict produces the following important consequences for the parties: (a) equalizing (putting parties on an equal footing); (b) attention-arresting (focusing attention on real problems); and (c) solidifying (increasing internal cohesion). Conflict also contributes to preventing social stagnation, encouraging growth and change, stimulating innovation, and generally producing consequences that are system-enhancing and strain reducing

(Bercovitch, 1984). The development of conflict in oil and gas operation in coastal areas is almost inevitable.

2. Conflict Resolution

Conflict is a disagreement or clash between ideas, principles, or people regarding a specific issue. Conflict resolution is the process of resolving a conflict regarding various issues such as the environmental impact of an industrial project. For instance, this paper deals with a conflict regarding environmental impact of oil and gas operations.

Conflict resolution is composed of the following steps (see Figure 1):

1) Conflict site location.
2) Conflict identification.
3) Conflict assessment.
4) Resolution strategy proposal.
5) Strategy implementation.
6) Evaluation.
7) Monitoring.

The following issues should be considered when dealing with oil and gas operations (see Figure 2):

1) Social issues.
2) Environmental issues.
3) Economic issues.

The social issues consist of impacts on:

1) Population health.
2) Population safety.
3) Community sustainability.

The environmental issues cope with impacts on:

1) Species.
2) Land quality.
3) Sea quality.

4) Water quality.
5) Air quality.
6) Ecosystems.
7) Environment sustainability.

The economic issues involve impacts on:

1) Commodity access.
2) Agriculture.
3) Fishery.
4) Industry.
5) Recreation and tourism.
6) Local economy.
7) Regional economy.
8) National economy.
9) International economy.
10) Economy sustainability.

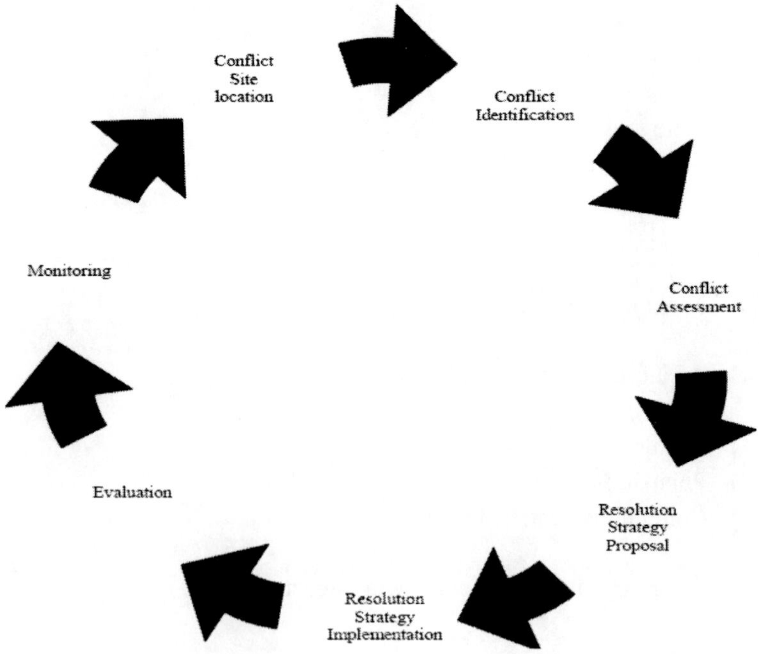

Figure 1. Conflict resolution steps.

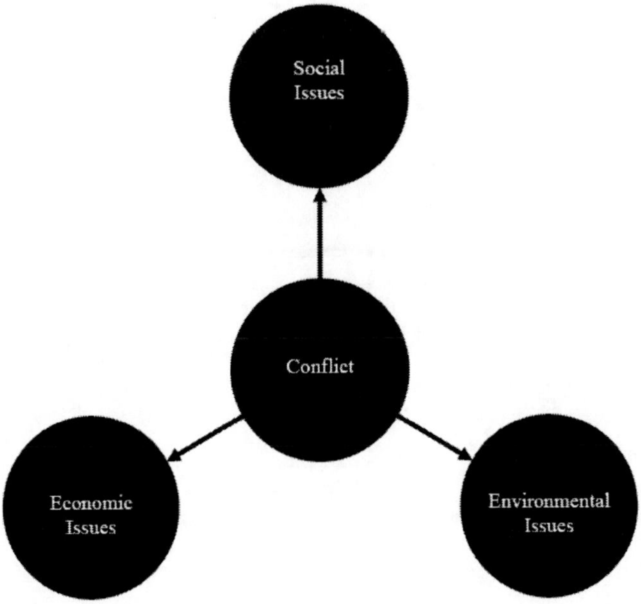

Figure 2. Conflict issues.

3. Description of Study Area

Cape Breton Island is located along the east coast of Canada and is part of the Maritime Province of Nova Scotia. The Island's culture and history has been diverse, with members of the Cape Breton community being of French, English, Scottish and Mi'kmaq descent. Cape Breton has a long history of cultural, recreational and economic connections to the coastal and marine environment. Because of this, many residents feel an entitled to the resource and a sense of ownership towards marine resources and space.

The economy of Cape Breton is primarily supported by resource-based industries such as fishing, forestry, mining, oil and gas. From 1990 to 2003, employment has risen nationally by 20.3 % and in Nova Scotia by 12.07 %. In Cape Breton, however, employment fell by 2.4 % over this same period (Figure 3). Canada's labour force grew by 19.7 percent and in Nova Scotia by 11.1 % since 1990. The labour force contracted in Cape Breton by 2.5 %.

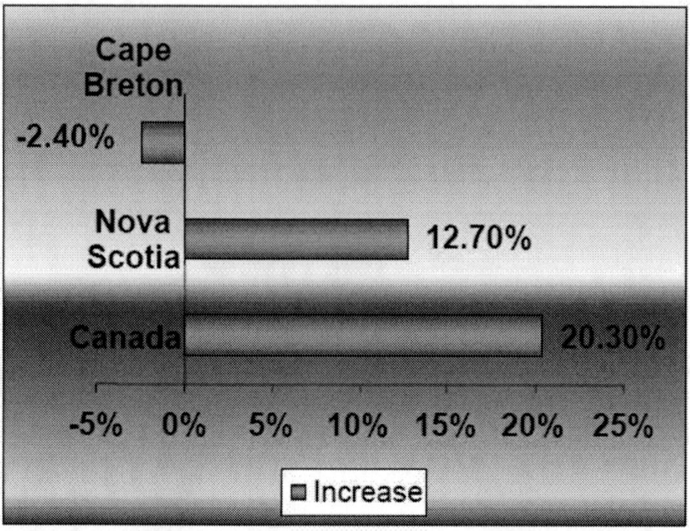

Figure 3. Comparative employment situation in Cape Breton (1990-2003) (Labour Market Review, 2004).

Year 2003 annual average employment in the **Forestry, Fishing, Mining, Oil and Gas** industry group was back to the 2001 level of 2,400 after a decline of 1,200 in 2002. This group accounted for 12.3 % of Cape Breton's employed in 1991 (Labour Market Review, 2004, Cape Breton). Declining fish stocks and the permanent closure of steel plant and coalmines have been severe blows to the local economy.

Today, Cape Breton is recognised as one of the poorest regions in Canada with unemployment estimated at 16.2 % (Labour Market Review, 2004, Cape Breton). Out migration linked to a lack of employment opportunities on the island has surged in the past few years. In an effort to address the high unemployment rate on the island, Cape Breton has been looking towards efforts that would diversify its economy. The tourism industry on the island is now the leading economic contributor. There is also the potential for natural gas development off the coast of Cape Breton, which will enhance the economy of the island as well as provide jobs to many of the local population. This potential development and the resolution of conflict it has generated is the prime focus of this paper.

4. Potential Areas of Conflict

On July 1, 1998, Hunt Oil Co. was issued exploration licenses (no: 2364 and 2365) off the northeast coast of Cape Breton, Nova Scotia. These licenses cover an area of land totaling 580,470 hectares. On July 1, 1999, Corridor Resources was issued exploration licence (no: 2368) off the western coast of Cape Breton. This licence covers an area of 247,020 hectares (Province of Nova Scotia, 2001).

Currently there are 59 exploration licenses worth $1.5 billion issued off of Nova Scotia (Province of Nova Scotia, 2001). Licenses issued off Cape Breton run from the shoreline to the offshore and have been the source of much public outcry. The proximity of the areas to the coastline has raised many concerns among residents and organizations with some interest groups vocally calling for the exploration licenses to be rescinded. In response to public protest, on October 20, 2000, the Canada-Nova Scotia Offshore Petroleum Board (CNSOPB) received a joint directive from the provincial and federal energy ministers to conduct a public review on the effects of potential oil and gas exploration and drilling, within exploration licenses 2364, 2365 and 2368.

The public review began in September and October 2001 where the Public Review Commission conducted issue identification sessions in 8 communities throughout Cape Breton and Prince Edward Island to hear comments and concerns of over 200 people (Public Review Commission, 2002). Official public hearings began on January 10, 2002 in Sydney, Cape Breton and moved to Montague, PEI on January 25, 2002 (Public Review Commission, 2002). The public sessions and hearings were open to all who wished to attend and there was an option of making verbal or written submissions to the Commission to voice concerns. In total over 130 submissions were made (Erskine, 2002). Many people criticize the public review as they feel it is a cosmetic process put in place by a government who has already decided to proceed with exploration without due consideration of the concerns of the public. The public review was also announced in a short time before a provincial election, which caused reinforced the existing high level of skepticism over how much public concerns would be considered.

No single root cause can be given for the present Cape Breton situation, illustrating the complexity of coastal conflicts. The various stakeholders provide different reasons. Established resource users (e.g. fishermen, tourism operators) see a potential threat to their livelihood from coastal oil and gas development. Government cites misinformation as the root of this particular problem. Environmental groups and some local residents see the long-term environmental integrity of coastal Cape Breton as being in danger from hydrocarbon development. Environmental groups have also expressed concern that an environmental impact assessment (EIA) was not triggered at the exploration license phase, considering the sensitivity of Cape Breton's marine environment. They feel that a strategic EIA, conducted by CNSOPB, was not sufficient in outlining potential impacts of oil and gas development on the marine environment. Industry is acting from economic interest and views the current conflict in Cape Breton as a delay to development.

5. Stakeholder Situation and Analysis of Conflicts

Stakeholder describes individuals, groups, or organizations that have an interest in the project and can mobilize resources to affect its outcome in some way. Most precise definition of stakeholder is 'individuals and organizations who are actively involved in the project, or whose interests may be positively or negatively affected as a result of project execution or successful project completion' (Smith, 2000). Stakeholder analysis provides opportunity and some basic measures that can help uncover and remove barriers. This includes information for additional tools that could prove valuable for the project team: a communication management plan, a risk management plan, and possibly a set of project requirements. A well-designed project will not only clarify key stakeholder roles, but will define as much as possible who participates when. Not all stakeholders need to be involved in all aspects of the project in all life cycle phases. Identifying relevant stakeholders is a challenging task. It is always difficult to ensure adequate representation of some important stakeholder groups due to cultural, religious, gender, economic or other factors during the project implementation.

For the purpose of the present study, the analysis of following stakeholders that interact in the oil and gas development in Cape Breton project area has been taken into account (Figure 4):

1) The Canada-Nova Scotia Offshore Petroleum Board (CNSOPB);
2) The Nova Scotia Petroleum Directorate (an agency of the provincial Department of Natural Resources);
3) Provincial Government: Nova Scotia Petroleum Directorate;
4) Fishing industry;

5) Non-governmental environmental agency (Ecology Action Centre);
6) Local residents (non-consumptive resource users);
7) Oil and Gas Industry (Corridor Resources).

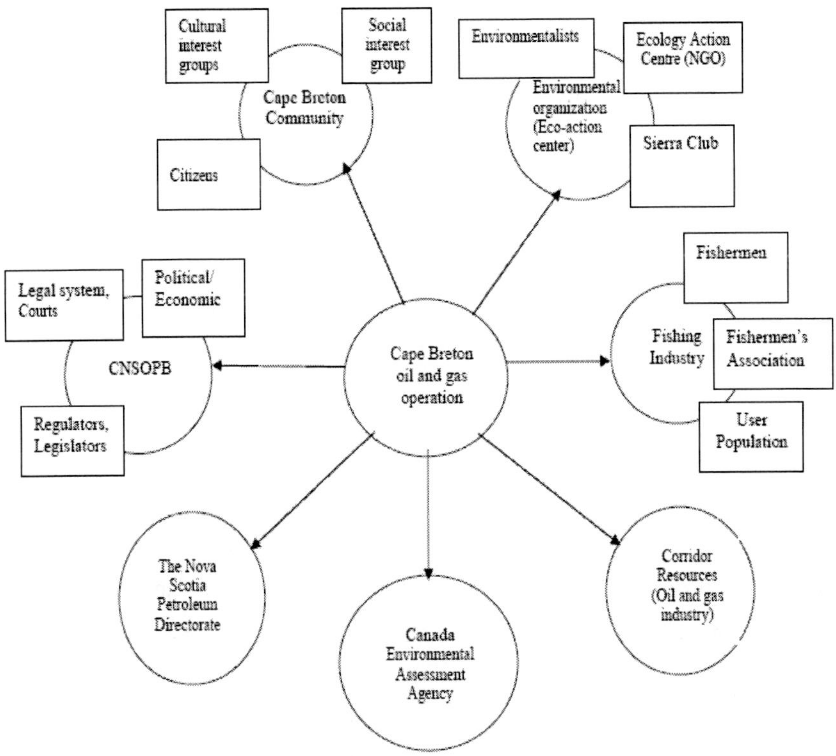

Figure 4. Stakeholder analysis: oil and gas operation in Cape Breton.

5.1. Canada Nova Scotia Offshore Petroleum Board (CNSOPB)

The Canada Nova Scotia Offshore Petroleum is an entity of the Federal and Provincial government that "regulates petroleum activities offshore Nova Scotia in an efficient, fair and competent manner" (CNSOPB). CNSOPB plays a double role as a regulator and promoter of the oil and gas industry in regards to the issuance of licenses for offshore exploration and development and is also responsible for ensuring that the environment is protected. Many view this duality of roles by the CNSOPB, as both the regulator and promoter of the Oil and Gas industry, as a conflict of interest.

The Board's licensing process is very complex and it is based on the Canadian Petroleum Resources Act (CNSOPB, 2001). During the nomination phase and prior to the Call for Bids, the Board along with Federal and Provincial government departments determine whether the identified areas occur within environmentally sensitive and fishing areas. There has been concerned raised by other stakeholders within the Cape Breton coastal areas that this processed outlined above was not adhered to by the CNSOPB. There is a need for work authorization for each phase of the offshore oil and gas lifecycle come form the Board before the commencement of any work, such as: seismic testing; exploration wells; and production activity, after the company has provided the Board with a comprehensive environmental impact statement and an environmental protection plan. The Board is the key authority to suspend any activities at any stage during the life cycle of oil and gas development if it detects any irregularities or violations in applying its regulations.

5.2. Canada Environment Assessments Agency (CEAA)

The Canadian Environmental Assessment Agency regulated under the Canada Environmental Assessment Act (CEAA) is a federal agency "that promotes environmental assessments as a planning tool to protect and sustain a healthy environment" (CEAA, 2001). Under CEAA, environmental impact assessments (EIA) can only be triggered under federal departments, of which the CNSOPB is, and although the CNSOPB's Accord Act states that it supersedes all other legislation, CNSOPB has agreed to follow and abide by CEAA regulations in reference to EIAs.

EIAs under CEAA can only be triggered with the identification of a project (CEAA, section 5 (1)). Project in the act refers to "any proposed construction, operation, modification, decommissioning, abandonment or other undertakings in relation to that physical work" (CEAA, section 2(1)). Thus the CNSOPB does not require EIAs from the proponent prior to the issuance of an exploratory license and again holding of a license does not give the company authorization to conduct any type of activity/project in the offshore. However under CEAA a screening will be most likely be required for seismic testing and exploratory drilling, with a comprehensive EIA required for production drilling. A screening process evaluates the magnitude of the anticipated impacts of a project and also determines whether there is a need to approve, modify or halt a proposed project. According to CEAA, 2001 Comprehensive studies are usually carried out for larger or valued ecosystem sensitive projects (CEAA, 2001).

5.3. Provincial Government: Nova Scotia Petroleum Directorate

According to the members of the Petroleum Directorate there is no conflict in Cape Breton except lack of public education. The assumption is people who reside in Cape Breton do not really welcome the petroleum industry as they think as an external invasion in their coastal areas. Lack of effort from the media in educating the general public of their overall benefits from coastal development creates lots of ambiguities. According to provincial government, the actual conflict, between two extreme views, one opposed to development and one in favour, which, in their opinion, is not the view of the general public. For example, the trade union (labourers) is in favour of development as it means employment and at the public review many squared off with fishermen who are opposed to oil and gas development.

According to the Petroleum Directorate decision should be based without considering all users and sectors (fishing, tourism, subsea mining, shipping, submarine cables, oil and gas etc.) as they think the ocean space off Cape Breton as an area of equal opportunity to all users and sectors. The need to create public awareness and educating peoples are the key to achieve a balance for the proper use of space. The public review is seen as a positive step by the provincial government as it allowed for the communication between user groups and government in a democratic fashion."Seizing the Opportunity: Nova Scotia's Energy Strategy" encourages exploration of Nova Scotia's natural resources. The provincial government acts on behalf of the citizens to promote responsible exploration and development of the oil and gas industry and as steward of the public good must proceed in the best interest of environmental integrity, health, safety and social objectives of the province (Province of Nova Scotia, 2001).

The offshore oil and gas operations can contribute in creating a strong and competitive economy and can effectively serve to diversify the economy of a relatively economically depressed portion of the province for Nova Scotia. Many people suspect the Cape Breton projects will also fail to employ a significant number of residents from the experience of Sable Offshore Energy Project (SOEP) as it failed to provide sufficient jobs among the local residents. A major objective of the provincial government is to maximize Nova Scotia's employment in the oil and gas sector (Province of Nova Scotia, 2001) and

Cape Breton development can reach that objective. The provincial government has complied monthly employment figures for SOEP reporting that in September 2001, 926 jobs were created in Nova Scotia, and 74% of new hires were residents of Nova Scotia (Anonymous, 2001).

The Petroleum Directorate recognizes natural gas in Cape Breton will at least provide employment that can deter out-migration of youth. Out-migration is a problem that is obvious in the –7.6 % population decline in Cape Breton from 1996 to 2001 (Statistics Canada, 2002). In 1996 6,045 out of 117,849 people were employed in manufacturing and construction (Statistics Canada, 2002), a figure the provincial government feels it can raise though oil and gas development. According to the Petroleum Directorate there is an ample opportunity for Nova Scotia to provide hydrocarbon products to a global market exists and Cape Breton can play a significant role in that context. Environmental groups and residents of Cape Breton have attacked the petroleum industry and government citing environmental threats as a need to prohibit development in nearshore areas. Many see a threat to the fishery and tourism industry that they feel cannot co-exist with large-scale hydrocarbon developments.

Since much of the threat to coastal zone off Cape Breton from hydrocarbon development is minimal but the federal and provincial regulatory controls over the petroleum industry are insufficient to protect the marine environment are unfounded. Rigs in operation are under strict regulatory controls. For example there is a zero discharge regulation, meaning that even rainwater running off rigs is collected and does not enter the marine environment. The industry as being under sufficient environmental regulations and say there may be more of an environmental threat from fishing where dragging is equivalent to clear cutting and bilge water from boats is often dumped directly into the ocean. The positioning of rigs is also not actually as close to the shoreline as has been reported in the media and should not impact tourism operations. Rigs will be visible but only on the horizon.

5.4. Position of the Fisheries

The conflict between the fisheries and the oil and gas industry with respect to the exploitation of common ocean space is, in nature, a classic marine conflict. Fishing is an industry, both traditional and modern in nature, employing and sustaining a large number of people and sustaining their livelihood. The concern is, therefore, on the oil and gas industry to adjust with this primordial oceanic activity in order to pave the way for a possible co-existence. Among the stakeholders and the interest groups involved in the current conflict, the primacy of the fishing industry is established by certain irrefutable facts like its long history, predominance over other industries, capacity to expand and sustain the livelihood of a number of people in the region. It is estimated that over 20,000 people are engaged in the industry at present (Hughes, 2000).

The success of the fishing industry is, in effect, testified by its ability to withstand the onslaught of the 'moratorium on ground fish industry (mostly cod) in 1992, and still survive and thrive on other catches i.e., lobster and snow crab. Both these industries are still very active and have created a balancing effect on the local economy (Nature Pulse, 2001). Besides, lobster and snow crab, mackerel, redfish, herring, grey sole, founder, and silver hake are being harvested and/or processed in the community. According to Fisheries and Oceans Canada (2002) the following few other factors further add to the importance of the industry in the region. Firstly, with the increase in the commercial value of snow crab, the fishery is gradually expanding to cover all fishing grounds along the west coast of Cape Breton; Secondly, the fishery in the area is likely to attain more importance in the coming years due to the potentialities for capturing foreign markets for under-utilized species; Thirdly, there is the non-traditional aspect of the fishing industry i.e., the growing development of aquaculture with the potentialities for creating job opportunities for the local communities;

Therefore, being the direct and traditional user of the ocean with long established interests in the sector, the fishers view the incoming oil and gas industry in the area as an encroachment on their occupational domain. The fishers view the conflict from three angles: competition for space, unforeseen economic effects of oil and gas industry on fishery and ecological impacts. Competition for space becomes more acute as the licensed areas for oil and gas industry fall within the inshore waters of Nova Scotia where the impacts on fisheries are more adverse than on those in the offshore areas (Maritime Fishermen's Union, 2000). The possible effects in the area would therefore be: loss of access to fishing grounds due to new installations, gear damage from the debris, reduced marketability of products as a result of contamination etc. Osborne Burke, President of the North of Smokey Fishers Association remarked, "The oil and gas companies do not care about Cape Breton. Fishermen, women and their children who may wish to continue to generate a sustainable income many years after the Oil and Gas industry have left the area" (Maritime Fishermen's Union, 2000). The fishers see oil and gas as being capital-intensive rather than labour-intensive and hence its capacity for limited employment. Also, in their own domain, they had to sacrifice quotas and place voluntary limits on bottom dragging gear. Thus, in a submission on the socio-economic profile of the Snow Crab Fishery, it has been shown that the sector earns millions of dollars annually, thus facilitating not only local employment but providing as well tax to governments at all levels – municipal, provincial and federal (Morris, 2002). Some of the adverse effects on fishing due to oil and gas activities with serious economic repercussions are identified in one of the submissions as: loss of fishing areas during the short-term exploration phases; loss of fishing areas over the long-term. Should exploration prove successful and production begin; damage to gear and vessels during the short-term exploration phases; damage to gear and vessel during the production phases; damage to the fishery from such factors as oil spills, emission from gas blow outs, discharge of produced water, other debris; and finally loss of long-term income producing capability resulting from an accidental occurrence of any kind related to oil and gas activities (Morris, 2002). However, what is of most concern to the fishers is that the licensed areas for oil and gas are in the midst of spawning, nursery and migratory areas for lobsters, salmon, recovery ground fish species, snow crab and are home to whales and dolphins (Mac Kinnon, 2002). Thus, any works or undertaking that may result in alteration or disruption of fish habitat is to be prohibited. The department of Fisheries and Ocean also expressed their concerns about the possible risks of exploration on fisheries (Sierra Club of Canada Cape Breton Island, 2002).

5.5. Environmental Organization: The Ecology Action Centre

The position of the environmentalists towards the oil and gas industry is one of avowed opposition to its activities. The conflict arises between the two debatable themes *Development vs. Conservation*, thereby attaining at times few philosophical and moral overtones for justifying denial of access to common marine environments by both. The conflict between the two has the potential to become real in nature i.e. a seismic survey vessel cutting loose a lobster trap. Viewing the oil and gas activity as a threat to economically important industries already existing in the area, the environmentalists support the fishery's view that the benefits of oil and gas are uncertain and if any, they would be short-term in nature and at enormous cultural, social and environmental costs. It is to be noted that the environmentalist position has been a very rigid one as various prominent organizations in the field like the Ecology Action Centre, and the Sierra Club have echoed their voice against oil and gas in a similar tone. Starting from the point that the licensed areas are highly diverse, productive and sensitive areas harbouring fishes, marine mammals, reptiles, endangered species and as well a focal point of biomass of many important groundfish species, the environmentalists see enormous risk to the ecosystems of the area from oil and gas activities. Some even see the risks from a much wider perspective like the issue of climate change and unavoidable greenhouse emissions from development (May, 2002).

A look at certain elements would clearly reveal the united position of the environmentalists against the oil and gas industry. They are enumerated as follows:

1) The environmentalists question the decision to carry on oil and gas activities in the nearshore areas with grave concern and limited knowledge in the field. The *Oceans Act* aims to protect the fisheries from over-exploitation and demands the application of the precautionary principle. Environmentalists want a "precautionary approach to marine development. In particular, the shallow, semi-enclosed and ice covered physical features of the area need to be taken into due consideration.
2) Most environmentalists are of the opinion that the licenses were issued to the oil industries in the absence of full and through review of environmental impacts and identification of sensitive zones, and hence the process is flawed and devoid of ecological concerns and economic imperatives of the people who make their living on the oceans.
3) The two permitted and routine petroleum activities of concern are seismic blasting and discharges from exploratory drilling. However, impacts of oil and gas activities on marine biota are not well known. Both laboratory and field experiments, apart from being costly, have serious limitations. This is particularly true with respect to a better understanding of the impact of seismic blasting. Thus, in submissions to the Public Review Commission, there is concern over significant gaps in knowledge of the impacts of seismic. The industry experts at the hearing eventually agreed with the opponent's statements that research is inadequate on many aspects of oil and gas impacts (especially for seismics) on the marine environment (Dressel, 2002).

4) The environmentalists are also averse to the idea that EIAs are to be project oriented – a factor that inhibits such assessment in many stages of oil activities.
5) Environmentalists see the violation of human rights in the process of issuing exploration licenses, as there had been no consultation with First Nations, fishermen or local communities.
6) The environmentalists are equally concerned with the adverse effects that oil and gas industry might have on the flourishing tourist industry in the region.

It should be noted that while there is the united position of the fisheries against the oil and gas industry, demanding a total moratorium on oil and gas activities in the region following George's Bank, there is still a segment within the fishing industry, however minor it may be, that still considers that oil and gas industry activities and the fishing industry can co-exist. As the submission by the Area 18 Crab Fishermen's Association stated, "We are also sensitive to the possibility of a finding that may be in favor of exploration and as such, we are urging you to recommend that a protective compensation package be put in place that would replace lost income resulting from accidents during the exploration and/or production related activities" (Morris, 2002).

5.6. Local Residents of Cape Breton

There appears to be a split in opinion on whether oil and gas exploration should proceed off Cape Breton's coast. This is an information gap as public opinion polls regarding the matter could not be obtained and no actual percentages of people for or against development can be given. The dispute has positioned some fishermen and environmentalists against hundreds of unemployed pipe fitters, tradesmen and labourers. Proponents see oil and gas related jobs as a necessity in light of closures of Cape Breton's steel and coal mines resulting in high unemployment. The environmentalists are in complete opposition but most of them do not consist of actual Cape Bretoners. There is a cultural divide between the tight community of Cape Breton and non-local groups. With high unemployment a valid concern and the cohesiveness of the Cape Breton community, many people do not want to oppose oil and gas development as it means a job for their neighbour.

There is a history of failed enterprise and poor approaches to resource development in Cape Breton that makes many residents sceptical of industry and government. People need to be more convinced that natural gas exploration and extraction will economically benefit the community and that most of the revenue will not go to an American company (Hunt Oil) or even to the smaller Halifax based Corridor Resources. There are strong ties to the ocean in Cape Breton as many generations of people have relied on marine resources for their livelihood. There is a sense of ownership of the marine environment and residents do not want to give away their resources without tangible benefit.

5.7. Corridor Resources-Local Oil and Gas Industry

Corridor Resources is a local oil and gas company that came into existence in 1995. The company focuses both on terrestrial and offshore development entirely within Eastern Canada: Atlantic Provinces and Quebec. The area that Corridor Resources has identified for possible exploration is approximately 15 km from the coast of Cape Breton, suggesting that the presence of rigs should not have a negative impact of the aesthetics views of the area. In other words the rigs will not be very noticeable from the shore and they were aware of the possible impacts that exploration could have on the fishing industry in the area, and very early in the process tried to undertake consultative meetings with the Cape Breton fishing community. However the fishers were frightened by the environmental groups opposed to oil and gas development, refused to sit with them.

According to Corridor Resources the main conflict was between environmental groups and those in support of development. These organizations had used the first opportunity available to them to get the media to take on their cause. The best way of intercepting the project was to do it at the earliest possible phase. This was what the environmentalist groups were doing by expressing their concern over the CNSOPB's licensing process. During the exploratory phase, there would be little benefits for the Cape Breton community. Long-term benefits would only be felt if a deposit large enough for production was found and even so, such benefits would take a while to be felt in the community. With regards to possible environmental impacts of oil and gas development, Corridor Resources is aware that they must address environmental impacts of oil and gas development. Due to the proximity of the license to the shore and the importance of fishing to the area, the company has stated that if allowed to undertake exploratory drilling in the area, mitigative measures will be put in place to ensure minimal environmental impacts. Some of these include a zero tolerance policy for discharge of waste into the marine environment. However, it was felt that gas found in the area would be sweet gas, compared to the highly polluting sour gas in Alberta, and as such should not pose much of an environmental threat to the area. Corridor Resources did feel that oil and gas development and fisheries would be able to eventually co-exist. However this could only happen when the oil and gas and fishing industries agree to start talking.

6. Discussions

It is apparent that there is conflict among the direct users (fishers) and indirect users (environmentalist) over the potential impacts of oil and gas development along the coastal area of Cape Breton. Most resource users in the stakeholder analysis had voiced concern over the impacts of oil and gas development, it should be stated that they did so in varying degrees. The environmentalists were adamant in voicing that a moratorium for oil and gas development should be in force for coastal Cape Breton. The fishers, who in part mirrored this sentiment, in some instances voiced willingness to attempt to work with the oil and gas industry. The comments from the residents seem to be mixed, with the majority more concerned with the economic long-term impacts of oil and gas, while the provincial government appeared to, at least in theory, be supportive of development in the area.

According to Cicin-Sain (1992) the conflict in the area seems to be revolving around (1) competition for ocean space (between the fishing and oil and gas); (2) operational impacts (pollution impacts from oil and gas on fishing, community health and the tourism industry; and (3) ecosystem impacts (impact of oil and gas on fish habitat and spawning grounds and impacts on the integrity of the marine environment). It should be noted that all these concerns arise from potential or imagined impacts of oil and gas development. To date there has been little or no activity in the nearshore area of Cape Breton, which has resulted in any of these impacts (Cicin-Sain, 1992). It should also be noted that there appears to be conflict in the area between government (federal and provincial) and environmental groups as well as the community of Cape Breton. The environmental groups have indicated that they were not informed about the possible oil and gas development off coastal Cape Breton during at the Call for Bids, thereby implying either that the government deliberately failed to inform them, or that the licensing process of the CNSOPB is flawed. Many environmental groups have also articulated their concerns about the CNSOPB acting as both a regulator and promoter of the Oil and Gas industry, which they see as a conflict of interest. The community of Cape Breton on the other hand has a long history of distrust of government and industry, which they feel in the past have exploited them. These concerns, compared to the aforementioned ones are actual and have occurred in the past or are presently happening.

The government and to some extent the industry have indicated that much of this conflict arose primarily due to misinformation of the impacts of development, expressed in the media, lack of public education and lack of communication between the resource users and stakeholders. For example, a documentary entitled "Battle At Our Shores" about the impending oil and gas development off Cape Breton showed oilrigs extremely close to the beach. This is incorrect for the proposed development projects where rigs would be visible from the shoreline but only on the horizon. Many people believed this information was correct for Cape Breton and it stirred emotional reaction.

It is obvious that the oil and gas industry, particularly Hunt Oil International and Corridor Resources, are in favour of oil and gas development off Cape Breton. However they too, like the environmentalists see the application of the precautionary approach as a means of ensuring the strong regulation of oil and gas development. Like that of sustainable development, different stakeholders have misused the precautionary approach/principle as a means of promoting their own interest. Jordan and O'Riordan (1999) indicated that at its core the precautionary principle should act in advance of science to protect the environment from incurring harm. In an attempt of achieving this, risk avoidance should be seen as a means of ensuring the implementation of this principle.

In many respects the Public Review hearings proved to be beneficial to the process, as it enabled communication between resource users who prior to the hearings had no communication with each other. Thus during these hearings the fishers were able to express their concerns about the impacts of oil and gas industry to the fishery such as: destruction of fishing grounds and gear; high mortality of fish eggs and larvae during seismic testing; and the possibility of tainting of fish. The oil and gas industry for their part were able to voice the mitigative measures that were put in place to ensure minimal environmental impact such as a zero discharge policy and not conducting seismic testing in spawning areas. The Public Review also dispelled many myths of oil and gas development that were being endorsed by the media. In many respects much of the information in the media appeared to have been skewed towards the environmental groups who, as has been previously mentioned are against any type of oil and gas development off Cape Breton. As a result of this, and rightly so, the fishers and more so the community of Cape Breton felt that oil and gas development off their coastline was a threat to their and the children's livelihood.

In traditional societies, turning to village leaders or elders may well be the most effective way to address disputes in the coastal zone (Johannes, 1978). In developed Western societies, coastal conflicts are often resolved through the efforts of neutral mediators who assist coastal managers in bringing the parties together to negotiate a solution (Cicin-Sain and Knecht, 1998). Bercovitch (1984) states that a successful resolution process involves changes in behaviour and perceptions of the parties involved.

Cicin-Sain and Knecht (1998) suggest that an effective Integrated Coastal Management (ICM) conflict resolution approach should have three important elements: (1) efforts to understand the roots, causes and consequences of coastal and marine conflicts through conflict 'mapping' studies; (2) an established and transparent process for making decisions about the conflicts; and (3) the capability to adopt and implement measures to remedy injuries or damage to particular coastal and ocean users from coastal development or from the actions of other coastal and ocean users.

There were also concerns raised by environmental and fisheries groups about the lack of an environmental impact assessment conducted at the exploratory phase. Although this concern may in fact be a result of a lack of understanding of the CNSOPB's policy in that EIAs are required prior to any work, CEAA may want to reconsider the process for triggering EIAs. It is apparent that the strategic EIAs that are conducted by the CNSOPB are not sufficient as they were unable to properly identify fragile and spawning areas that should have been barred from any oil and gas activities.

One of the major criticisms found in the entire licensing process was the lack of communication between, government, industry, the Cape Breton community, the fishing community and environmental groups. It is important that some form of communication exist between the fishing and oil and gas industry. Presently, it appears that the oil and gas industry is in direct competition for space with the fishing industry that has existed in the area for many years. Meetings with fishers and the community can serve as public education sessions where the potential environmental and economic risks of oil and gas development can be made. Based on the above discussions and stakeholder analysis, the following recommendation is made in order to resolve the potential conflict in Cape Breton oil and gas operations.

7. Recommendations

Conflict resolution methods vary greatly according to the socio-economic and political contexts of particular situations. It also results when parties are able to abandon power-orienting strategies that often treat only symptoms and embrace a participatory approach towards management. The most common means pertaining to conflict resolution in the marine environment in the literature involves introduction of Integrated Coastal Management (ICM). This holds true for the Cape Breton experience. ICM provides a mechanism in which all parties, by being involved in the management of the area, would be in a better position to successfully address concerns before they appear irresolvable, as is presently seen in Cape Breton. Like in other institutional solutions, public participation plays a major role in ICM as managers recognise that the users have valuable insights regarding both in the management and economic initiatives of a management program.

In Cape Breton, there will always be a need to closely allocate and manage competing uses in a given area for the optimum long-term benefit of the people of the Island. It is, however, hoped, that the integration of economic and environmental concerns will always be given careful consideration by all stakeholders at all stages of development, including elected political representatives. Moreover, there is a need for the political will and courage on the part of the elected members of government to support the agencies involved in coastal management and to uphold legislation and policies. It is imperative for the public to be more proactive and aware of activities and decisions that are made, which have a grave effect on the coastal resources and the social and economic well being of the Island. ICM represents the current phase in development planning where the interactions between land and sea are recognized as having a major influence on how coastal systems respond to human activities. ICM also recognizes the influence of policies, investment strategies and development plans governing human activities have on the ability of coastal systems to sustain human development needs and aspirations. Based on the ICM principles, the following recommendations have been suggested for conflict resolution in oil and gas activities in Cape Breton.

(1) Presently, it appears that the oil and gas industry is in direct competition for space with the fishing industry that has existed in the area for many years. It is important that some form of communication exist between the fishing and oil and gas industry. Meetings with fishers and the community can serve as public education sessions

where the potential environmental and economic risks of oil and gas development can be made. The community is then able to make decisions and be a part of the integrated coastal management of the area. The need for the CNSOPB to review its licensing process is therefore recommended, by ensuring that the public plays a more proactive role in any future oil and gas activities occurring in Nova Scotia's offshore.

(2) It appears that the CNSOPB has a very sectoral approach towards management in their obvious *faux pas* in involving other important stakeholders and governmental agencies. Implementation of ICM in Cape Breton includes co-ordination, consultation and planning process designed to derive general agreement, and providing for conflict resolution, when required. This process facilitates all stakeholder involvement in community-based decision-making, and supported as required by appropriate levels and agencies of government. Implementation of a comprehensive program of education, training, health, security, research and communication is the heart of the conflict resolution. An improved system of access to information and to regulatory processes including monitoring, evaluation and feedback is prerequisite for Cape Breton case.

(3) Oil and gas projects result in a number of direct and indirect local economic and community impacts including immigration, inflation, social equity/parity, local food, housing, water supply, schools and recreational pursuits. Working with local communities and assessing potential effects in advance of construction can provide opportunities for government investments in upgrading local infrastructure to promote good neighbour practices through community compensation and investment. Creation of business opportunities are methods through which oil and gas development in Cape Breton can contribute to local economic development and the growth of sustainable enterprises.

(4) The development of conflict in the marine areas is almost inevitable and it is in no way being implied that the establishment of ICM would resolve disputes between and among stake holders, rather ICM provides a mechanism in which all parties, by being involved in the management of the area, would be in a better position to successfully address concerns before they appear irresolvable, as is presently seen in Cape Breton.

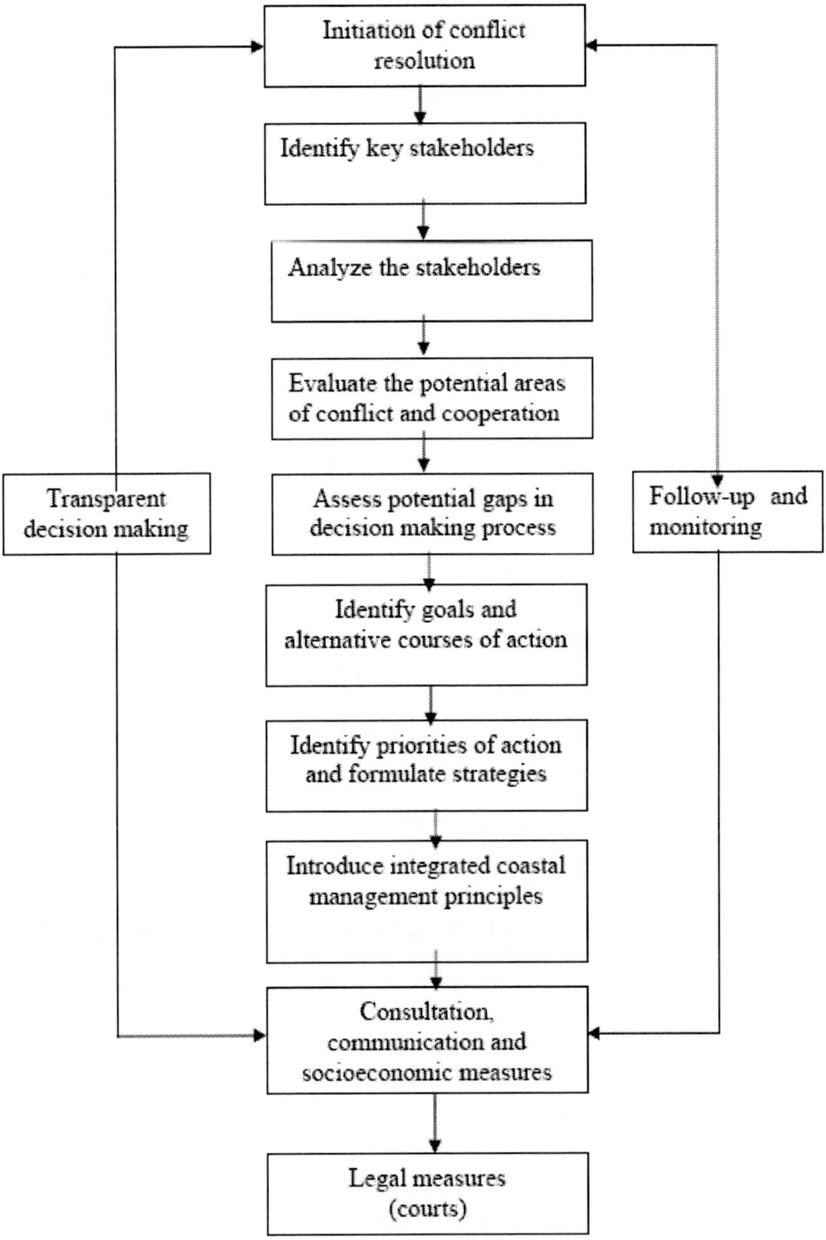

Figure 5. Proposed model for conflict resolution in Cape Breton oil and gas operations.

8. Proposed Approach for Conflict Resolution

A model for conflict management based on the ICM principles is proposed for Cape Breton case (Figure 5). All stakeholders and potential areas of conflict and cooperation are identified. Any gaps found during the decision making process are assessed and strategies and alternative courses of actions are taken to address them. An integrated policy involving participation of all stakeholders which includes transparent decision making, consultation, and socio economics aspects are given due consideration before going to the legal systems to resolve conflict.

Conclusion

A detailed analysis of a complex conflict was carried out regarding oil and gas activities in Cape Breton, Nova Scotia. Various stakeholders and their role are described in order to resolve the possible conflicts. The community based conflict resolution model was proposed based on the Integrated Coastal Management (ICM) principles as an approach for possible conflict resolution in oil and gas operation in Cape Breton.

Acknowledgment

The authors would like to thank the Atlantic Canada Opportunities Agency (ACOA) for funding this project under the Atlantic Innovation Fund (AIF).

References

Anonymous (2001) *Nova Scotia employment figures – Sable Offshore Energy project*, Nova Scotia Petroleum Directorate, Halifax, NS.

Atria Engineering Hydraulics Inc. (1995) *Coastal Conservation Project, Initial Development of a Coastal Zone Management Framework for St. Lucia, Pre Feasibility Study Phase 1 Final Project Report*, Government of St. Lucia, Ministry of Planning Development and Environment, Castries, St. Lucia.

Bercovitch, J. (1984) *Social conflicts and third parties: Strategies of conflict resolution*, Westview Press, Colorado.

Brown, N.A. (1997) *Devolution of authority over the management of natural resources: The Soufriere Marine Management Area, St. Lucia*. Case Study Series by the Caribbean Natural Resource Institute.

Canada Environment Assessment Agency (CEAA) (2001) *Environment assessment,* Internet document, <http://www.ceaa.gc.ca/0009/index_e.htm>.

Canada Environmental Assessment Agency (CEAA) (1992) *Canada Environmental Assessment Act,* Government of Canada.

Canada Nova Scotia Petroleum Board (CNSOPB) (2001) *The Offshore Licensing Process*, Internet document, <http//:www.cnsopb.ns.ca/Rightmgt/licencingProcess.html>.

Cicin-Sain, B. (1992) 'Multiple use conflicts and their resolution: Toward a comparative research agenda', In. *Ocean Management in Global Change*, ed. P. Fabbri: Elsevier Science Publishers Ltd., England.

Cicin-Sain, B. and Knecht, R.K. (1998) *Integrated coastal and ocean management: Concepts and principles*, Island Press, Washington D.C.

De Bono, E. (1985) *Conflicts: A better way to resolve them*, London, Harrap Limited.

Department of Fisheries and Oceans (2002) Internet document, <http://www.ifmt.nf.ca/mi/aac>.

Dressel, H. (2002) 'Big oil moves in on Cape Breton Shorelines', *Dateline*, February 22, 2002.

Erskine, B. (2002) 'Cape Breton oil exploration still on hold', *Chronicle Herald*, Wednesday, March 27. Halifax, N.S.

Fisher, R., Urg, W. and Patton, B. (1992) *Getting to yes: Negotiating an agreement without giving in*, rev ed., Hutchinson, Boston.

Hughes, S. (2000) *Halifax Chronicle Herald*, Halifax, Nova Scotia.

Johannes, R.E. (1978) 'Traditional marine conservation methods in Oceania and their demise', *Annual Review of Ecological Systems*. Vol. 9, pp. 349-364.

Johnson, P.J. and Duinker, P.N. (1993) *Beyond dispute: Collaborative approaches to resolving natural resource and environmental conflicts*, School of Forestry, Lakehead University Thunder Bay, Ontario.

Lindbaek, J. (2004) Keynote speech, UN Global Compact Symposium in New York: *Strengthening Conflict-Sensitive Business Practices in Vulnerable and Conflict-Affected States*.

MacKinnon L. (2002) Submission to Public Hearing Process: Exclusion Zones: Sensitive ecosystems should be protected from a potential oil and gas disaster, Internet document, <http://www.publicreview.ns.ca>.

Maritime Fishermen's Union (2000) 'Fishermen say no to oil/gas exploration on Cape Breton's coastline'. Internet document, <http:/www.sierraclub.ca/national/...breton-alert/mfu-pr-rel-09-00.html>.

May E., (2002) *Submission to the Public Review Commission on behalf of the Sierra Club of Canada*, Internet document, <http://www.publicreview.ns.ca>.

Morris, J.P. (2002) Submission to Public Hearing Process: Compensation for fishermen in the event of an oil and gas disaster, Internet document, <http://www.publicreview.ns.ca>.

Nature Pulse (2001) *Fishermen protest seismic testing*.

Province of Nova Scotia (2001) *Seizing the opportunity: Nova Scotia's energy strategy*, Government of Nova Scotia, Halifax, N.S.

Public Review Commission (2002) *Public Review Commission*, Internet document, <http://www.publicreview.ns.ca/>.

Saint Vincent de Paul Society (2000) Internet document, <http://highlander.cbnet.ns.ca/~vdawson/social.htm>.

Service Canada (2004) *Labour Market Review*, Cape Breton, Canada.

Sierra Club of Canada Cape Breton Island (2002) *Action Alert*, Internet document, <http//www.sierraclub.ca/national/action-alert/cape-breton-alert/html>.

Smith, L.W. (2000) 'Project clarity through stakeholder analysis', CrossTalk - The Journal of Defense Software Engineering.

Statistics Canada (2002) Statistical profile of Canadian communities. Internet document. <http://ceps.statcan.ca/english/Profil/PlaceSearchForm1.cfm>

Suman, D. (2001) 'Case studies of coastal conflicts: Comparative U.S./European experiences', *Ocean and Coastal Management*, 44(1-2), pp. 1-13.

Yanez-Arancibia (1999) 'Terms of reference towards coastal management and sustainable development in Latin America: introduction to Special Issue on progress and experiences', *Ocean and Coastal Management*, Vol. 42, pp. 77-104.

In: Sustainable Petroleum Operations
Editor: Rafiq Islam
ISBN 978-1-61324-157-8

Chapter 5

Technological Analysis and Quantitative Assessment of Oil and Gas Development on the Scotian Shelf, Canada

***M. I. Khan*[1] *and M. R. Islam*[2]**
[1]Institute for Sustainable Energy, Environment, and Economy
ES602-2500 University Dr, NW, University of Calgary, Calgary, AB, Canada
Email: Ibrahim.Khan@ucalgary.ca
[2]Faculty of Engineering, Dalhousie University-Sexton Campus
D510-1360 Barrington Street, Halifax, NS, B3J 2X4, Canada

Abstract

Offshore oil and gas development is a high technology intensive operation. In this paper, an overview of oil and gas development on the Scotian Shelf of Nova Scotia, Canada, and its technological details are examined. For the purpose of risk and environmental impact assessment, the detailed technological phases and scale of oil and gas operations are analysed. The technology phases are categorized into four phases: seismic exploration, exploratory drilling and installation of structures, production, and decommissioning of the project. The technological details and scale of operation involved for each of these phases varies significantly. Each technological phase is critically analysed and it is found that each phase produces/releases numerous wastes, none of which is remedied completely. Besides, each phase of the operation is inherently inefficient and wastefully in energy and mass. This paper also studies, the typical candidate locations and jurisdictional situation for oil and gas project lifecycles. It provides quantitative information of offshore oil and gas development activities in the Scotian Shelf and uses this as a basis for risk and environmental impact assessment. This study will be helpful to engineering and policy makers and can bridge the gap between the two.

Keywords: offshore oil and gas; quantitative assessment; technological analysis; life cycle assessment; Scotian Shelf.

1. INTRODUCTION

Offshore oil and gas operations draw huge public attention due to their environmental consequences. All stages of offshore oil and gas operations (OOGOs) generate a variety of solids, liquids and gaseous wastes (Currie and Isaacs, 2005; Wenger et al., 2004; Khan and Islam, 2003a; Veil, 2002; Groot, 1996; Wiese et al., 2001; Rezende et al., 2002; Holdway, 2002). However, high market demand and technological advancement in OOGOs spread all around the world and even in remote and deeper oceans (Wenger et al., 2004; Pinder, 2001). On the Scotian Shelf, explorations were started as early as 1960 (Figure 2). Considering the pattern of usage by the petroleum industry, the Scotian shelf can be characterized as one of periodic high interest (DFO, 2003).

Not withstanding this intermittent approach, the fact that this shelf is under exploitation makes it important to consider the potential impacts and associated risks of oil and gas activities. Different technological activities in each phase of oil and gas development have different impacts and risk on marine ecosystems (Groot, 1996; Khan and Islam, 2003a; 2003b; 2004; 2005a; 2005b; 2005c; 2007 and 2008; Khan et al., 2005). To understand the risk scale and impact levels in issue, the technological information to be considered can be divided into the four phases of the typical offshore oil and gas extraction process of the Scotian shelf project lifecycle. These phases are: 1) seismic exploration; 2) exploratory drilling and installation of structures; 3) production; 4) decommissioning of the structure. The technological activities of each phase have different impacts and risk on the marine ecosystem (Holdway, 2002; Groot, 1996; Managia et al., 2005).

On the Scotian shelf, extensive exploration was done between 1967 and 2004 (Figure 1). Approximately, 385,000 km of seismic tracks have been made using 2-D technology and 25, 000 km^2 with 3-D (Figure 2). The total survey cost was $168 million between 1967 and 1997 (DFO, 2003). Drilling operations were undertaken after each of the peaks in seismic survey activity. One hundred and ninety-nine wells have been drilled on the Scotian shelf with an expenditure close to $5.0 billion (Figure 2). Most of the wells are categorized as exploratory, and only a few of them are development wells. There have been two offshore development projects: Cohasset/Panuke Project (CoPan, now abandoned), and Sable Island Offshore Energy Project (SIOEP), with three sites already developed and at least two more underway (Curran et al., 2005). The project lifetime is expected to be at least 20 years.

Overall, most of the available information in the field of OOGO relates to engineering or technological advancement such as the well reported work of Peters and Fowler (2002); Sadiq et al. (2005); Pinder (2001), Islam and Wellington (2001), Malik and Islam (2000). In addition, limited information for non-technical persons, such as policy makers, legislators, managers and stakeholders are reported (Khan and Islam, 2003a). This article provides quantitative information on offshore oil and gas development activities on the Scotian shelf as a basis for risk and environmental impact studies. The study considers these issues in light of the four phases that a viable project goes through. In terms of benefit, study would be useful to legislators, managers, government agencies and departments, and international organizations as they seek to formulate policies, regulatory frameworks and guidelines with respect to offshore operations. The information given here can also be help bridge the gap between science and policy in the regulation and management offshore operations. The study below is organized in terms of the technological challenges at the four consecutive stages of a project cycle, that is, seismic explorations, explorations drilling, production, and decommissioning respectively. Concrete experience on the Scotian shelf is used for illustration. As well, jurisdiction over resource ownership, regulations and environmental protection responsibility are considered.

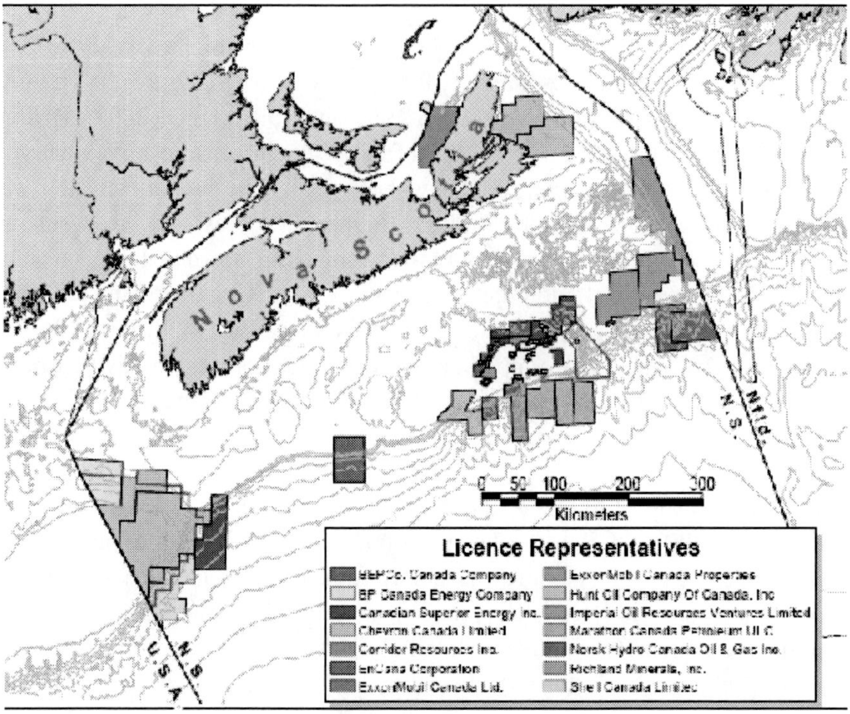

Figure 1. Active exploration licenses in the Nova Scotia offshore area including Sable Island.. Source: CNSOPB (2005).

2. Seismic Exploration

2.1. Technology Details

The first seismic operation in the Scotian shelf was initiated in 1960. By 2004, 400034.33 km of 2-D seismic tracks were completed. Since 1985, there have been 3-D seismic exploration undertaken, and 29511.86km² has been completed (Figure 1). Marine seismic exploration uses the behaviour of induced sound waves to map sub-seafloor longitudinal sections (2D) or features (3D). Different seismic activities and their related effects are shown in Figure 2, the four cylinders of which represent the four major activities in this area which are connected sequentially. They are towing airguns, blasting airguns, towing hydrophones and chase boating. The arrows with numbers represent types of input and output or impact that the activities have.

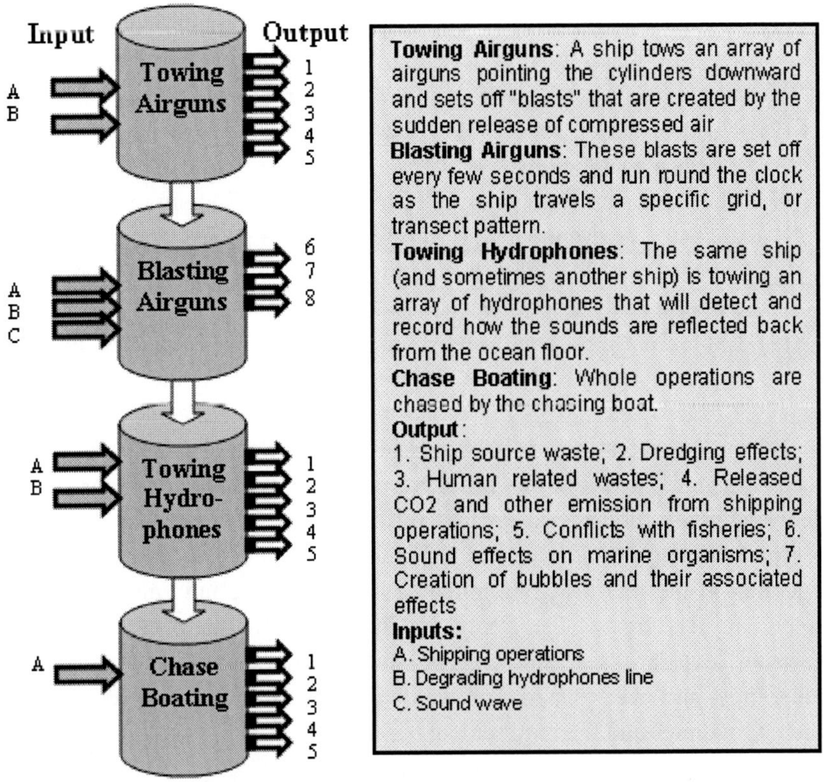

Figure 2. Schematic Seismic Activities in offshore exploration involving towing airguns, blasting airguns, towing hydrophones and chase boat operations.

The 'towing airguns' activity is done by a boat, which tows an array of streamers with several airguns attached below the water surface (Figure 3). These release high-pressured air every 12 seconds, creating a sound of pulse generally 225dB to produce seismic waves in the water (Figure 4). The airguns are focused at the seafloor, reflected by sub-seafloor features and received by hydrophones attached every 12.5 m along streamers up to 6 km long. A typical run lines are 100 m to 1 km apart, as shown in Figure 3. The 3-D survey needs more closely spaced lines that yield highly resolved spatial structures, with the need for interpolation. These operations are escorted by 'chase boats' to protect the dragging unit (Figure 3) from other boat traffic.

The following activities take place during a typical seismic exploration in an offshore operation.

(1) A ship tows an array of airguns aiming the cylinders downward and sets off "blasts" that are created by the sudden release of compressed air.
(2) These blasts are set off every few seconds around the clock as the ship travels a specific grid.
(3) The same ship tows an array of hydrophones. Sometimes another ship tows the hydrophones. Hydrophones detects and records how the sounds are reflected back from the ocean floor. Different types of rock and other substrate reflect sound differently.
(4) Explosives were once used when underwater seismic exploration first began, but they have been replaced by compressed air

During the operation, air bubbles rise to the surface that interfere with the received signal. To avoid this problem, multiple airguns of varying sizes are used. Generally, noises are created by other types of sound in the marine environment. Figure 4 shows the different possible types of sounds that are generated in the marine environment, as well as their intermittent and local effects. In this figure, the frequency of sound and their pressure density spectrum level are also shown. The seismic study involves very complex technology, and every seismic project is different from another depending on where it operates and the requirements of the survey.

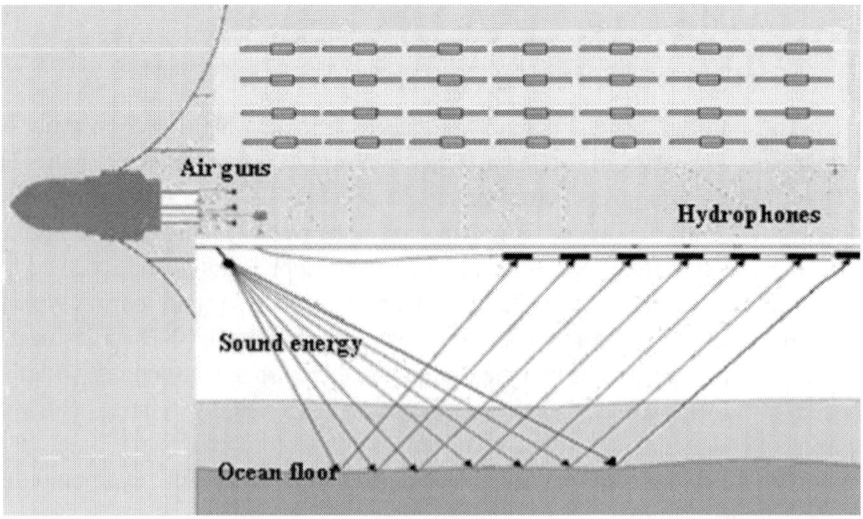

Figure 3. Pictorial view of a typical seismic operation in the marine environment, where a ship tows an array of airguns and hydrophones.

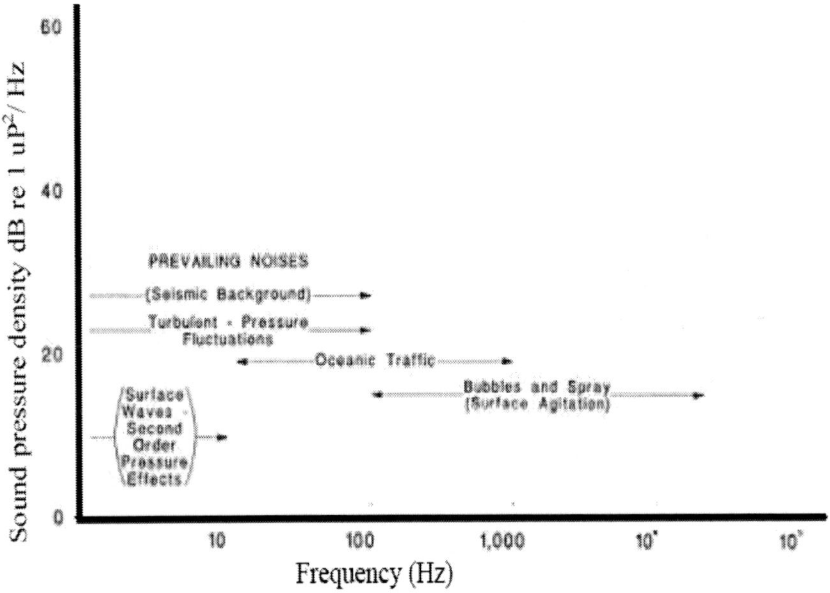

Figure 4. Frequency level and sound pressure density spectrum of different types of sounds in the marine environment.(According to Wenz (1962); Kndudsen et al.(1944))

2.2. Technology Scale

Seismic surveys generally operate for a period of 5 to 20 days. During operations, activities run 24 hours/day. The area for each survey varies. Current leases in Nova Scotian waters range from 10,000 to 40,000 km^2, and survey areas may only cover a proportion of the lease area. A survey takes the form of a grid of line which may range from 20 to 1000 km in length, but typically about 100 km^2. Total seismic survey in the Scotian shelf is presented in the Figure 5 (data source: CNSOPB, 2005).

From 1960 to 2004, there were 400,034.33 km of 2D seismic tracking. The peak survey period in Scotian Shelf was 1999 to 2000. After the introduction of 3D seismic technology in 1985, there have been a total of 29,511.86 km^2 also done on the Scotian shelf. Van Daessel (1991) reported that the effects of 2-D and 3-D are similar and both physically damage, disturb and even kill marine organisms (Jepson et. al, 2003). The total cost of the seismic survey was about \$168 million between 1967 and 1997 (DFO, 2003).

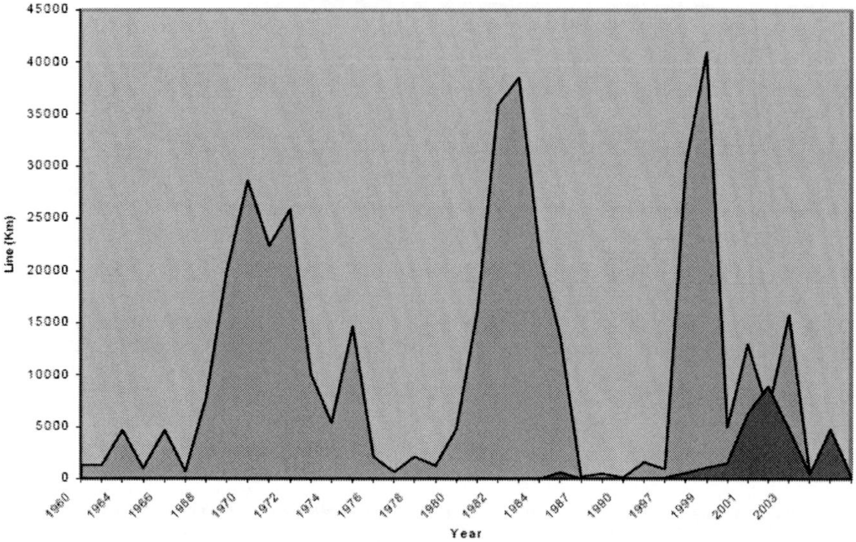

Figure 5. Amount of 2-D and 3-D seismic (line Km) shooting on the Scotian shelf.

3. Exploratory Drilling and Installation of Structures

3.1. Technology Details

3.1.1. Structures

Seismic surveys provide a general idea of the presence of hydrocarbons based on the structure and properties of rocks. However, exploratory drilling confirms the presence of hydrocarbons. Production drilling takes place if the project site is determined to be economically feasible based on the quality and quantity of the hydrocarbon reservoir. There are basic differences between exploratory and production drilling. In exploratory drilling, the diameter of a drillhole is smaller than in production drilling. As a result, different drilling-cuts are generated (Groot, 1996).

For a drilling operation, suitable rigs are selected based upon environmental and geotechnical conditions. There are submersible and semi-submersible rigs, jackup rigs, as well as drill ships available. For example, submersible rigs are used in shallow water (up to 25 m) and jack-up rigs operate in deeper water, such as 20-125 m. They consist of self-contained legs lowered to contact the seabed. The mat-supported rig is used in soft seafloor sediments. Generally, submersibles and drill ships are used in deeper water. Security and risk to a structure have become important issues after accidents involving world's largest, 40-storey-tall offshore oil platform, in Brazil on March 20, 2001. Quantitative risk assessment is also now applied to offshore installations (Ramsay, 1994).

3.1.2. Drilling

Regardless of the type of rig used, the drilling operations are similar. The main task of a drill rig is performed by the hosting, circulating and rotation system, backed-up by the pressure-control equipment. A common drilling operation is pictorially shown in Figure 6. In this picture, the drill bits, drill mud input and screening of drill cuttings are clearly shown. A drill bit is attached at the end portion of a drill pipe. Motorized equipment rotates the drill pipe to make it cut into rocks. During drilling, many pumps and prime movers circulate drilling fluids from tanks through a standpipe into the drill pipe and drill collar to the bit. The muds flow out of the annulus above the blowout preventer over the shale shaker (a screen to remove formation cutting), and back into the mud tanks (Figure 6).

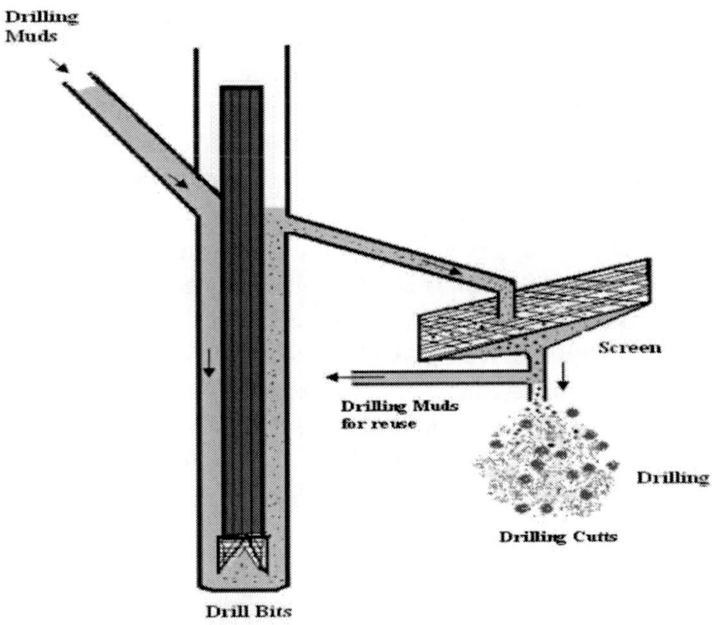

Figure 6. A pictorial representation of drilling that includes drill-bits, shaker-screen and input of drilling-muds.

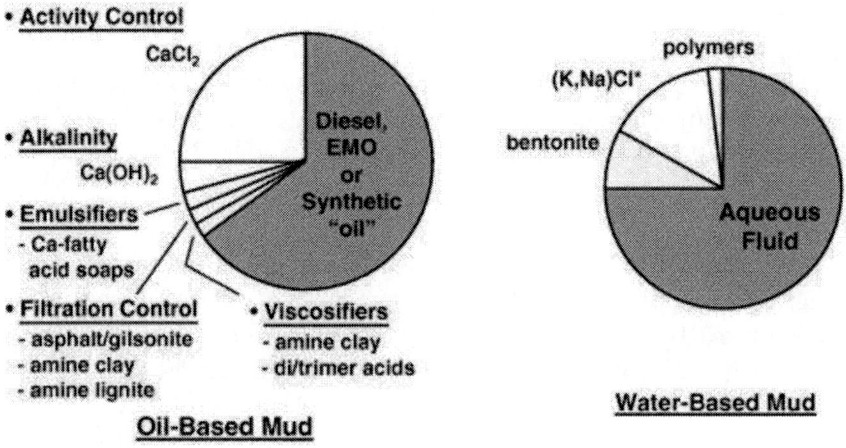

Figure 7. Composition of oil-based and water-based drilling muds (in weight %, excluding weighting agents) used in offshore oil and gas operations (Source: Wenger et al., 2004).

In a drill hole, a casing is needed when the drill pipe goes deeper. Generally, for the first 1000 m, casing is not required. For greater depths, the dull string is removed and a pipe-the "well casing"- is inserted into the well. As drilling progresses, the well is lined with additional casing both to prevent rock from crumbling into the hole and to contain any high-pressure gases and liquids. On top of the casing, the blowout preventer is added to close the well in the event of uncontrollable pressures.

Mainly water-based, oil-based and synthetic-based drilling muds are used in exploration and production drilling (Wenger et al., 2004; Khan and Islam, 2003a and 2007). The composition is ranged from a simple clay-water mixture to a complex blend of minerals chemically suspended in water and oil. The composition of drilling muds is shown in Table 1 (data source: Zwicker et al., (1983); Patin (1999) and Wenger et al., 2004). Water-based mud is composed of water and bentonite, and heavy minerals are also added for weight. Chemical additives are mixed in to stabilize the drilling fluids during use, and to reduce corrosion and bacterial activity (Table 1). Figure 7 shows the composition of oil-based and water-based drilling muds in percentage weight, excluding density control.

Table 1. Composition of drilling muds which are used in oil and gas exploration

Product	Composition	Concentration
Base fluid	Water, Bentonite clay (Sodium montmorillonite), Caustic soda (sodium hydroxide)	As needed
Additives	Lignosulfonate, Phosphates (sodium acid pyrophosphate and tetrasodiumpyrophosphate), Plant tains (predominant usage of quebracho), Lignite	1-2.7 kg/bbl
Density control	Barite (natural barium sulfate ore), Ferrophosphate ore, Calcite, Siderite, Hematite, and heavy metals.	0 – 317.5 kg/bbl
Fluid-loss control	Starch (corn and potato), Polyanionic cellulose polymer, Xanthum gum, Sodium carboxymethyle-cellulose, lignite.	<0.45-4.5 kg/bbl
Product	Composition	Concentration
Lost Circulation	Ground nut shells, Micas, Ground cellophane, Diatomacheous earth, Cottonseed hulls, Ground or shredded paper	0.9-13.6 kg/bbl
Corrosion and scale control	Sodium sulfite, Zinch chromate, Tall oil, Amines, Sodium hydroxide, Phosphates, Bactericides	0.11-2.7 kg/bbl
Solvents	Isoprophanol, Glycerol, Isobutanol, Ester alcohols, Diesel oil	----
Lubricant	Asphalts, Diesel oil, Fatty soaps, gilsonite, Glass beads, Rosin soap; Enhanced mineral oil; Synthetic oil (no aromatic content)	.09-2.7 kg/bbl

The oil-based muds are composed of mineral oils, barite and chemical additives (Table 1). Oil-based muds are used for deeper well sections and in cases where the well is drilled at an angle. In case of synthetic-based mud, mineral oil is replaced by oil like substances. This makes it more environmentally acceptable. Requirements as to appropriate drilling mud depend on well depth and the geological conditions of the drilling area.

3.2. Technology Scale

As noted already, drilling activities on the Scotian shelf were initiated in 1967. By 2004, 199 wells had been drilled, most of which are exploratory. Figure 8 shows the total number of wells drilled, and their depths in this study area. Out of 199 wells, 56 wells were drilled for production and 136 for exploration. In comparison to the North Sea and Gulf of Mexico, there were wells drilled (Managia et al., 2005). According to Groot (1996) in the North Sea, 1530 exploration, 997 appraisal and 2256 production wells were drilled. On the Scotian shelf, most well drilling periods were relatively shorter, i.e., less then three months. At present, the industry is actively drilling in the deeper slope waters. The peak drilling activities took place in the year 1970-79 (CNSOPB, 2005).

Drilling a well releases drilling muds, cuttings and produced waters into the marine environment. This remains an issue of concern (Khan and Islam, 2003a; Cranford et al. 2003; Rezende et al., 2002). In Figure 9, the systematic drilling activities are shown. Small arrows with numerical numbers represent different types of waste released into environment.

Drilling depths vary with location. A better way to understand the physical impact might be in terms of distance drilled (Groot, 1996). In Figure 8, distance drilled by each well is presented. Until 2004 a total 796,283 m distance was drilled. The average well depth on the Scotian shelf is about 4000 m below the sea floor (Figure 8). A well about this depth typically generates 1000-5000 m^3 of wastes (Groot, 1996; Patin, 1999). A production platform generally consists of 10-15 wells, which generates (12 x 5000 cubic meter) 60,000 m^3 or tons of wastes.

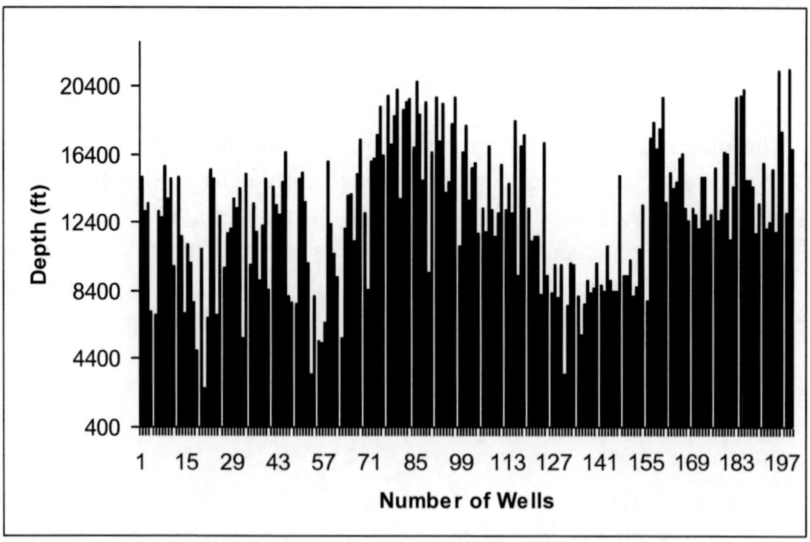

Figure 8. Number of wells drilled on the Scotian shelf, 1967 to 2004.

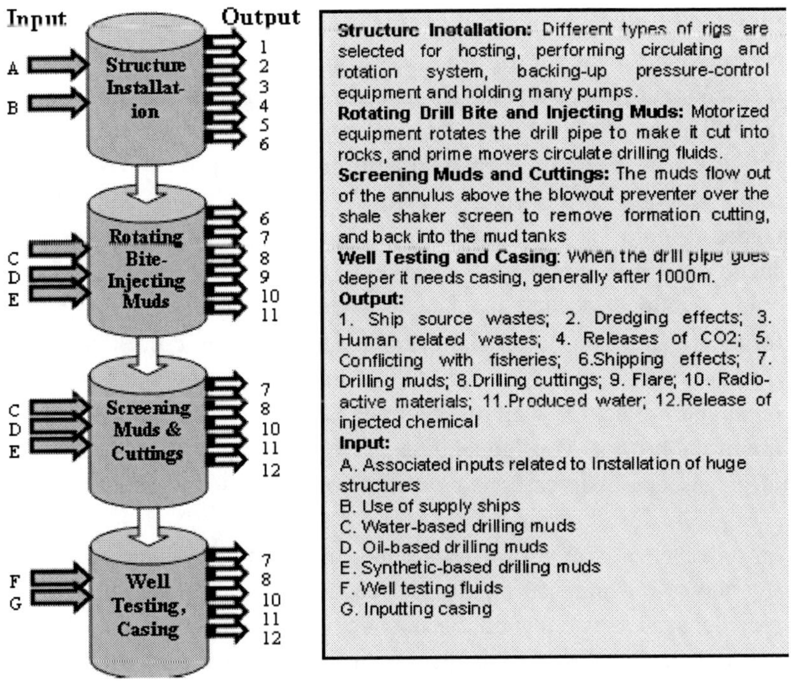

Figure 9. Systematic diagram of drilling activities and associated waste releases.

All rock removed from a well is deposited on the seabed. For an average well, this is equivalent to a column 5000 m deep, with a diameter of 90 cm at the surface, to about 20 cm at the bottom.

The volume of rock can range from 300 to 1200 cubic meters, and the volume of mud and cuttings combined can reach 3200 cubic meters from each exploratory well (estimates based on Groot (1996), Putin (1999)). However, it is unclear how long it takes for the cutting piles to become totally eroded by wave action and tidal currents (Groot, 1996; Kingston, 1990).

4. Production

4.1. Technology Details

The SIOEP of the Scotian shelf is the largest offshore gas project of its kind in North America today. The technology details described below focus on this project in particular. However, deviations are noted, given the size and scope of other offshore oil and gas projects that one is studying. This will simply provide an outline of what to expect in terms of the technological requirements of offshore oil and gas-extracting and processing units.

SIOEP, once it is completely finished, will consist of six platforms (one at each field). The Central Thebaud Platform Complex is the only manned platform, and the rest (Venture, North Triumph, Glenelg, and Alma) are unmanned (Figure 1). The unmanned platforms are called satellite platforms. Production platforms are the combination of jackets and associated equipment (top side facilities), which manage production. Many types of production platforms are found in the offshore, reflecting the specific needs of the field and hydrocarbon source. The "jacket" includes the pipes or casings, which surround and protect the well on the seabed. They are fixed to the seabed for the life of the project. Topside facilities have the equipment needed to capture liquid or gaseous hydrocarbons as well as crew quarters, generators, cranes, etc.

In order to get gas out of the ground, the natural pressure in a reservoir usually forces oil and gas to the surface. However, as reservoirs of liquid hydrocarbons, such as crude oil, become depleted, drillers inject water into the well below the level at which the hydrocarbons are found. Because oil is lighter than water, this pushes the oil up, and helps it flow to the surface. (Natural gas reservoirs normally produce their own pressure.)

4.1.1. Central Thebaud Platform 'Topside Facilities'

The Thebaud production and processing platform is the only permanently manned structure for the entire offshore portion of SIOEP. It actually consists of two platforms connected by a walkway. One of the platforms houses the Thebaud wells, while the other contains processing and production equipment for gas from Thebaud, as well as gas brought in from the satellite platforms, and living quarters (sleeps 40) and a helicopter landing pad. Steel jackets, meaning their legs reach the ocean floor, support both of these platforms.

The production complex's main function is the dehydration and separation of the produced gas, that is, to remove as much water from the gas as is technologically possible. Gas brought in is cooled by an inlet cooler, and separated in a three-phase inlet group separator into water, condensate (oil), and gas. Two of these separators are located on the central complex, with one equipped to handle low pressure in the future, when gas pressure in the wells decreases.

The combined gas stream is then fed into two triethylene glycol gas dehydration contractor trains whose function is to remove most of the water vapour from the gas. The triethylene glycol is then recovered by boiling the water off, and sending the triethylene glycol back to the contractor. The condensate, which was removed back at the inlet separators, is then combined and sent through a condensate coalescer and stripper to remove water. De-watered condensate is pumped back into the gas downstream of the triethylene glycol contractors. This final product of de-watered gas and condensate (oil) is sent through a pipeline to shore, and to the gas plant at Goldboro. Water vapour is vented into the air and any liquid water is sent to a water separation and treatment system, then discharged into the sea.

The central platform's power and water systems consist of multiple generators capable of running on natural gas or diesel fuel (only used for the start up of the platform or emergency power), and filtered seawater. Its fire protection and safety systems consist of a fixed extinguishment system, inert gas fire suppression, foam firefighting capabilities, seawater deluge, survival suits, lifeboats, temperature and pressure monitoring and control, and ventilation and pressurization detection systems.

4.1.2. Satellite Platform Facilities

The satellite platforms are designed to be unmanned and to contain as little processing equipment as possible. This enables most of the machinery to be centred on the Thebaud platform. Gas produced at the satellite platforms is separated in a three-phase group separator (water, gas, condensate). Produced water is treated in a hydrocyclone separator, which separates out more condensate using centrifugal force, and is then discharged overboard. The gas and condensate are then recombined and sent to the Thebaud complex (via subsea interfield pipelines) for further processing. Monoethylene glycol and corrosion inhibitors are injected into the flow to prevent the formation of hydrates in the gas and corrosion of the pipelines.

4.2. Technology Scale

Going from exploration to production is a major step in the development of an offshore reserve. The difference is mostly one of size: an offshore development has many of the same activities as a single exploratory well, presented in Figure 8 and discussed in section 3.1, but on a larger scale and for a longer period of time. Offshore "production" involves all the activities needed to actually develop an oil or gas reserve. It is more expensive and lasts longer than exploration (Figure 11). Unlike exploration, it requires the installation of permanent platforms at the wellhead. Given that the SIOEP is the largest project of its kind in North America today, the figures for scale will be higher than the average offshore oil and gas project in Nova Scotia. Figure 10 shows the production volume of SIOE project (Data source: CNSOPB, 2005). It shows the monthly average production of gas and associated water from January 2004 to September 2005.

The SIOEP is located 200km off of Nova Scotia, 10-40km north of the Scotian shelf. Drilling below the seafloor can go as deep as 5km. The platforms required for offshore oil and gas are separated by many kilometers.

For example, in SIOEP, the distance between the central Thebaud platform and the farthest satellite platform, Alma, is 50 kilometers, and the closest two platforms (Venture and South Venture) are separated by more than 5 kilometers. The amount of materials required to construct the production facilities is immense. For example, the SIOEP project used more than 900,000 meters of pipe, 109 million kg of steel, 53,000 meters of chrome tubing, and 160,000,000 kg of concrete. The SIOEP project is expected to yield over 100 billion m^3 of natural gas over the next 20-25 years.

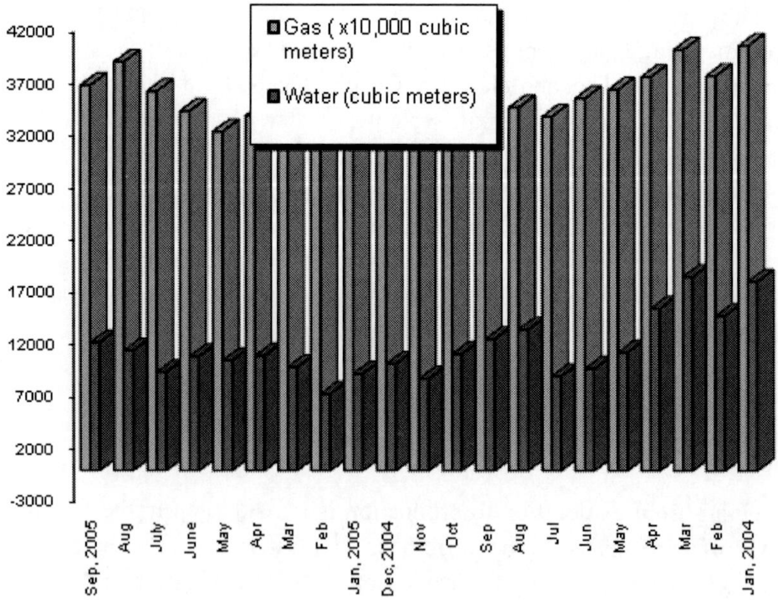

Figure 10. Average Monthly Production of raw gas (ten thousand cubic meters) and produced water (cubic meters) in the Sable Energy Project from Janurary 2004 to September 2005.

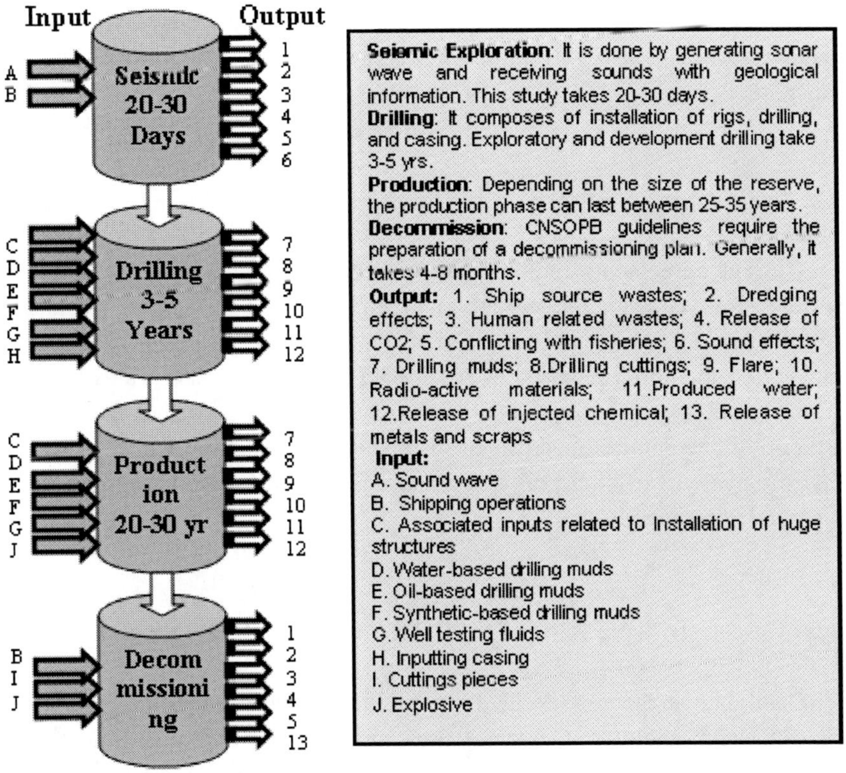

Figure 11. Life Cycle of Offshore Oil and gas development on the Scotian shelf.

5. DECOMMISSIONING

5.1. Technology Details

Decommissioning is the last step in offshore oil and gas development (Figure 11). It involves the removal of structures used for the operation, as well as any environmental remediation considered necessary (Schroeder and Love, 2004). Provincial regulations provide a description of the minimum of activity that must be undertaken at this stage. This entails clearing the seafloor of any material or equipment that could interfere with other commercial uses of the sea, preventing fluid from escaping from the well-bore, and cutting off all casing at a level below the so as not to be affected by ice scour.

In removing jack-up rigs, structures may require being cut into small manageable pieces. In 90% of cases, explosives are used. The structure pieces are then transported by barges to shore for disposal, reuse, or recycling. Disposal at sea is also an option when structures are deemed too difficult to transport.

5.2. Technology Scale

Current agreements require that all projects in waters up to 100 m be entirely removed with the exception of pipes. Those in deeper waters must have the top 55 m removed, and objects weighing up to 4000 tonnes must be transported to shore or buried at sea. Besides structures, the removal of drill cutting piles is currently under consideration. In Figure 11, the tentative duration of decommissioned projects is present. This figure, overall, shows the total life cycle of offshore oil and gas development. It also shows the wastes released from each phase.

6. Typical Candidate Locations

Candidate areas are defined by previously collected data, or data of adjacent land geology. Current leases suggest that there is potential for petroleum development in the Gulf of St. Lawrence, the Scotian shelf edge, and the Scotian shelf. The Bay of Fundy does not appear to be under consideration, presumably because of a lack of potential for deposits. The CNSOPB will not accept nominations for the area under moratorium on the Georges Bank, and the 'Sable Gully'. The depths associated with the continental shelf range from 50 to 100 m, with an average depth of 60 - 80 m. Particular sites within the Gulf of St. Lawrence are approximately 40 m.

7. Jurisdictional Situation

For all phases of petroleum development, a similar jurisdictional authority applies. Federal jurisdiction appears to take precedence over provincial in marine affairs, but any land-based activity is within provincial jurisdiction. Property rights are not clearly defined; thus, federal, provincial and First Nations titles are asserted over mineral rights.

Federal-provincial jurisdictional overlap is addressed by the formation of a joint provincial-federal body under the Canada-Nova Scotia Petroleum Resources Accord Implementation Act, 1987 (CNSOPB Act), although First Nations involvement has not been solicited to date. The CNSOPB exists as one regulatory body operating according to memorandum of understanding (MOU) with the federal Department of Fisheries and Oceans, Environment Canada, Nova Scotia Department of Transportation, Nova Scotia Department of Natural Resources, and the Nova Scotia Department of Environment.

Provincial guidelines for the operation of petroleum extraction initiatives are designated in the Nova Scotia Offshore Area Petroleum Geophysical Operations Regulations, Guidelines on Plans and Authorization Required for Development Projects, Guidelines on the issuance of exploration licenses, and offshore waste treatment Regulation guidelines. The United Nations Convention on the Law of the Sea (UNCLOS, 1982) defines Canada's international obligations in the decommissioning of offshore operations.

The CNSOPB executes environmental protection responsibility. The activities implicated in this fall under the domain of the Fisheries Act, the Oceans Act, and the Oceans Dumping Act, as well as protection required by provincial acts. Federal Authority to discharge the obligations has been vested in the CNSOPB. Thus, CNSOPB has to ensure observations of the Canadian Environmental Protection Act (CEPA) which is triggered respecting installation and production. Seismic and exploratory drilling, as well as site abandonment are, however, not activities to which CEPA applies, and CNSOPB has no obligation to enforce observance of CEPA in these cases.

CONCLUSIONS

The demands of the technological phases and the scale of oil and gas operations constitute the main bases for the environmental impact and risk assessment of an offshore development project. In this paper, the life cycle and quantitative assessment of offshore development of the Scotian shelf in terms of its detailed technological phases are analysed. It is found that every technological phase is responsible for the release of numerous solid, liquid and gaseous wastes. This finding suggests that there is a need for the development of alternative technology which is environmentally more acceptable in view of the likely adverse effects of this activity. This study provides an assessment of the current state of oil and gas development activities. It finds that there have been fewer amounts of seismic tracks on the Scotian shelf and wells drilled in comparison to other identical regions. Considering the multi-stakeholders' uses of the shelf this study should help to provide a comprehensive picture of OOGOs to all users and policy makers so as to guide them in environmental law-making and enforcement in this sector.

REFERENCES

CNSOPB, 2005. *Activity Reports*, Canada-Nova Scotia Offshore Petroleum Board, Halifax, Nova Scotia,Canada.

Currie, D.R. and Isaacs, L.R., 2005. "Impact of exploratory offshore drilling on benthic communities in the Minerva gas field, Port Campbell, Australia", *Marine Environmental Research*, Vol 59: 217–233.

Cranford, P.J., Gordon, D.C. Jr, Hannah, C.G., Loder, J.W., Milligan, T.G., Muschenheim, D. K. and Shen, Y., 2003. "Modelling potential effects of petroleum exploration drilling on northeastern Georges Bank scallop stocks", *Ecological Modelling,* Vol 166, No. 1-2: 19-39.

Daessel, J.P. van. 1991. "Marine mammals and seismic research in the Dutch Continental Shelf". *1st SPE Conference on health, safety and environment in oil and gas exploration and production*, 11-14 November, The Hague, The Netherlands, Vol 2: 573-577.

DFO, 2003. "State of the Eastern Scotian Shelf Ecosystem". *DFO Can. Sci. Advis*. Sec. Ecosystem Status Rep. 2003/04.

Groot, S.J.de, 1996. "Quantitative assessment of the development of the offshore oil and gas industry in the North Sea". *ICES Journal of Marine Sciences*, Vol. 53: 1045-1050.

Holdway, D.H., 2002. "The acute and chronic effects of wastes associated with offshore oil and gas production on temperate and tropical marine ecological processes", *Marine Pollution Bulletin*,Vol 44: 185–203.

Islam, M.R. and Wellington, S.L., 2001. "Past, Present, and Future Trends in Petroleum Research." *SPE paper no. 68799*, SPE Western Regional Meeting Held in Bakersfield, California, 26-30 March 2001.

Jepson, P.D., Arbelo, M. Deaville, R., Patterson, I.A.P., Castro, P., Baker, J.R., Degollada, E., Ross, H.M., Herráez, P. Pocknell, A.M., Rodríguez, F., Howie, F.E., Espinosa, A. Reid, R.J., Jaber, J.R., Martin, J. Cunningham, A.A. and Fernández, A., 2003. "Gas-bubble lesions in stranded cetaceans. Was sonar responsible for a spate of whale deaths after an Atlantic military exercise?" *Nature,* Vol 425: 575-576.

Khan, M.I. and Islam, M.R., 2005a. "Assessing the Sustainability of Technological Developments: An Alternative Approach of Selecting Indicators in the Case of Offshore Operations". *IMECE2005-82999. ASME Congress*, Orlando, Florida, Nov 5-11, 2005.

Khan, M.I. and Islam, M.R., 2005b. "Environmental Fate and Behaviour of Oil Discharges in Marine Ecosystem: Using Fugacity Model". *Third International Conference on Energy Research and Development (ICERD-3).* State of Kuwait, November 21-23, 2005.

Khan, M.I. and Islam, M.R., 2005c, "Environmental Modeling of oil discharges from produced water in the marine environment", *GC-225*, 33rd Annual General Conference of the Canadian Society for Civil Engineering, Toronto, Ontario, Canada, June 2-4, 2005 / 2-4 Jun 2005.

Khan, M.I, Zatzman, G. and Islam, M.R., 2005. "New Sustainability Criterion: Development of Single Sustainability Criterion as Applied in Developing Technologies". ICEC 2005. Amman, Jordan, 12 - 14 September 2005.

Khan, M.I. and Islam, M.R., 2004. "Assessing Environmental Fate and Behaviour of Oil Discharges in Marine Ecosystem: Using Fugacity Model". In: Proceedings of *the Offshore Oil and Gas Environmental Effects Monitoring Workshop: Approaches and Technologies*, Bedford Institute of Oceanography, May 26 - 30, 2003. Armsworthy, S.L., P.J. Cranford, and K. Lee (Eds.), Battelle Press, 145-165.

Khan, M.I., and Islam, M.R., 2003a. "Wastes Management in Offshore Oil and Gas: A Major Challenge in Integrated Coastal Zone Management". In: L. G. Luna (ed), *CARICOSTA 2003 - 1st International Conference on Integrated Coastal Zone Management ICZM*, University of Oriente, Santiago du Cuba, May 5-7, 2003.

Khan, M.I., and Islam, M.R., 2003b. "Ecosystem-Based Approaches to Offshore Oil and Gas Operation: An Alternative Environmental Management Technique", *SPE Paper: 84058*, Society of Petroleum Engineer Annual Technical Conference and Exhibition, SPE Conference, Denver, USA. October 6-8, 2003.

Khan, M.I. and Islam, M.R., 2007. rue Sustainability in Technological Development and Natural Resources Management. Nova Science Publishers, New York, USA: 381 pp.

Khan, M.I. and Islam, M.R., 2008. Handbook of Sustainable Oil and Gas Operations Management, Gulf Publish Company, USA: 461 pp.

Kingston, P.E., 1990. Impact of offshore oil production installations on the benthos of the North Sea. *ICES CM 1990/Mini*: 7. 6pp.

Kndudsen, V.O., Alford, R.S., and Emling, J.W., 1944. Survey underwater sound-ambient Noise (Report no. 3), Office of Scientific research and development, National Defense Research Committee, Division 6, Section 6.1., USA.

Malik, Q. M. and Islam, M.R., 2000. "CO2 injection in the Weyburn field of Canada: optimization of enhanced oil recovery and greenhouse gas storage with horizontal wells"., *SPE paper no. 59327*. SPE/DOE Improved Oil Recovery Symposium held in Tulsa, Oklahoma, 3-5 April 2000.

Managia, S., Opaluchb, J.J., Jinc, D., Grigalunasb, T.A., 2005. "Technological change and petroleum explorationin the GulfofMexico". *Energy Policy*, Vol 33: 619–632.

Patin, S., 1999. *Environmental impact of the offshore oil and gas industry.* EcoMonitor Publishing, East Northport, New York. 425 pp.

Peters, K.E. and Fowler, M.G., 2002. "Applications of petroleum geochemistry to exploration and reservoir management", *Organic Geochemistry*, Vol 33: 5–36.

Pinder, D., 2001. "Offshore oil and gas: global resource knowledge and technological change". *Ocean and Coastal Management,* Vol 44: 579-600.

Ramsay, C.G., Bolsover, A.J., Jones, R.H. and Medland, W.G., 1994. "Quantitative risk assessment applied to offshore process installations. Challenges after the piper alpha disaster", *Journal of Loss Prevention in the Process Industries*, Vol 7, No. 4: 295-304.

Rezende, C. E., Lacerda, L.D., Ovalle, A.R.C., Souza, C.M.M., Gobo, A.A.R., Santos, D.O., 2002. "The effect of an oil drilling operation on the trace metal concentrations in offshore bottom sediments of the Campos Basin oil field, SE Brazil". *Marine Pollution Bulletin*, Vol 44: 680–684.

Sadiq, R., Khan, F.I., and Veitch, B., 2005. "Evaluating offshore technologies for produced water management using GreenPro-I—a risk-based life cycle analysis for green and clean process selection and design", *Computers and Chemical Engineering,* Vol 29: 1023–1039.

Schroeder, D.M., and Love, M.S., 2004. "Ecological and political issues surrounding decommissioning of offshore oil facilities in the Southern California Bight", *Ocean and Coastal Management*, Vol 47: 21–48.

Veil, J.A., 2002. "Drilling Waste Management: past, present and future". *SPE paper no. 77388.* Annual Technical Conference and Exhibition, San Antonio, Texas, 29 Septermber-2 October. Waste Management Practices in the United States," prepared for the American Petroleum Institute, May 2002.

Wenger, L.M., Davis, C.L., Evensen, J.M., Gormly, J.R., and Mankiewicz, P.J., 2004. "Impact of modern deepwater drilling and testing fluids on geochemical evaluations", *Organic Geochemistry*, Vol 35 (2004) 1527–1536.

Wenz, G.M., 1962. Acoustic ambient noise in the ocean; spectra and sources, J. Acoust. Soc. Amer., Vol. 34.

Wiese, F. K., Montevecchi, W.A., Davoren, G.K., Hunttmann, F., Daimond, A.W., and Link, J., 2001. "Seabirds at Risk around offshore oil platforms in the North-Atlantic" , *Marine Pollution Bulletin*, Vol 42, No. 12 : 1285-1290.

Zwicker, S.L, J. Collins, J.T.E. Gilbert, J.E. Moore, and R. J. King. 1983. "Technical Guidelines for offshore oil and gas development". *In For the east-west environment and policy institute.* J.T.E., Gilbert (ed). Pp 49-104. Penn Well Books. Oklahoma City, Oklahoma.

In: Sustainable Petroleum Operations
Editor: Rafiq Islam

ISBN 978-1-61324-157-8

Chapter 6

SENSITIVITY ANALYSIS OF GENETIC ALGORITHM PARAMETERS IN GROUNDWATER MONITORING NETWORK OPTIMIZATION FOR PETROLEUM CONTAMINANT DETECTION

Abdolnabi Abdeh-Kolahchi*, Mysore G. Satish, Chefi Ketata and M. Rafiqul Islam
Department of Civil Engineering, Dalhousie University, Halifax, Nova Scotia, B3J-2X4, Canada

ABSTRACT

In order to detect petroleum contaminants in groundwater, it is vital to design and optimize the monitoring network properly and efficiently. So far, the best optimization techniques to solve this type of problems use genetic algorithms. The ability of Genetic Algorithm (GA) methods to identify optimal solutions in groundwater management optimization problems has been widely demonstrated. However, one of the main concerns in applying these methods is identifying suitable values for the GA parameters, especially population size, crossover probability, mutation probability, and usage of elitism. The values selected for these

* Corresponding author: E-maiL: kolahchi@dal.ca.

parameters have a significant impact on the objective function solution as well as the time taken to find that solution. Due to interaction of these parameters, the impact of each parameter depends on the values of the other parameters. This paper uses sensitivity analyses as a tool to identify the suitable values for GA parameters for a groundwater monitoring problem. Through these analyses, suitable parameter values are identified that cause the algorithm to perform very well in terms of finding final optimal solution.

Keywords: Genetic Algorithm parameters, sensitivity analysis, groundwater monitoring network design, petroleum contaminants.

1. INTRODUCTION

Genetic algorithms (GA) are based on the idea of modelling a search process based on natural evolution. It uses random elements to avoid local minima and uses terminology from biology and genetics. GAs gained a lot of popularity as a general-purpose optimization algorithm. Such techniques do not require computation of derivatives as used in more traditional optimization methods. In addition they are very popular in application of groundwater management. Unlike gradient-based methods, the GA uses a random search procedure inspired by biological evolution allowing only the fittest to survive and propagate to successive generations. The method starts with a number of possible solutions, which selected randomly, referred to as the initial population. Each of the possible solutions in the initial population is referred to as an individual, typically encoded as a binary string. For each individual, the objective function is evaluated. During the course of the research, new generations of individuals are reproduced from the old generations through random selection, crossover, and mutation based on certain probabilistic rules.

The use of the GA was first suggested by Holland (1975), who based his research on a survival-of-the-fittest rule. Since then, the GA has been used in many disciplines. At the core of the GA are three basic operators: selection, crossover, and mutation. GA procedures are detailed in Goldberg (1989) who presents a comprehensive introduction to the GA as does Davis (1991) who reviews many important applications of the GA. In addition, Haupt and Haupt (1998) show how to obtain a suitable solution to a genetic algorithm optimization problem.

GAs have been used for many water resources case studies. Applications include, but not limited to, water distribution system, reservoir system operation, groundwater remediation, and groundwater monitoring network design. Many of these models are nonlinear due to the behaviour of groundwater processes and management models which can solve by GA. GAs have shown themselves to be discovering good solutions with a reasonable amount of time and computational effort (Holland, 1992).

Given the random nature of any GA-based method, this study is concerned with the right selection of GA parameters such that the solution is faster and achievable. For example, using low mutation probability ignores the solution trap in dealing with local optimal values. GA parameters such as the size of the population, crossover probability, and mutation probability must be also chosen with care.

2. Problem Statement

An aquifer is contaminated by a source of pollution such as petroleum spills. An unspecified number of monitoring wells must be installed to monitor groundwater contamination movement and condition. Monitoring of groundwater contamination in a three-dimensional transient system requires a large number of monitoring locations associated with large implementation and operational costs. The optimal monitoring network design is based on the objective of maximizing the probability of tracking the contamination plume by determining sequential monitoring locations. A mathematical formula (objective function) based on monitoring network design is developed to identify and determine the number of monitoring wells, their location, their depth, and their time of implementation. The formulation of the optimization model is reduced from two objective functions to one (trace the migration of the plume). Then a genetic algorithm is used to solve for the optimal solution, i.e., to find the optimal locations of monitoring wells. The objective function, 34 constraints, and 64 variables are involved. Since there are 64 candidate wells to select for monitoring (i.e., 64 variables), there are 2^{64} or billions of possibilities. One of these possibilities will satisfy the objective function.

The success and performance of the GA depends on several parameters such as:

- Population size.
- Number of generations.
- Probabilities of crossover.
- Probability of mutation.

- Usage of elitism.

These parameters largely determine the success and efficiency of GA in solving a specific problem. Unfortunately, these parameters interact with each other in a complicated way. Researchers have been trying to understand the interaction of GA parameters, for example what value of mutation probability is suitable for a GA to achieve the optimal solution.

The interaction of these parameters is complex and difficult to achieve. This study uses sensitivity analysis to reach the suitable values for GA parameters for a groundwater monitoring network design.

It has been suggested that good GA performance requires the choice of high crossover and low mutation probabilities. Some researchers have suggested that choice of a mutation probability value which is inversely proportional to the number of variables (length of the string) would be enough to prevent the search from looking onto a local optimum (Carroll, 1996).

Deb and Agrawal (1998) investigated the effect of three parameters of genetic algorithm, mainly, population size, crossover probability rate, and mutation probability rate. The results showed that mutation and crossover-based approaches perform quite differently between simple problems and complex problems. In either approach, a correct population size is needed to achieve good performance.

Tongchim and Chongstitvatana (2002) present an adaptive algorithm that can adjust parameters of genetic algorithm according to the observed performance. The parameter adaptation occurs in parallel to the running of genetic algorithm. Four parameters are adjusted by the algorithm:

- Crossover operator.
- Mutation operator.
- Crossover rate.
- Mutation rate.

Two performance measurements are used to compare the proposed method with the algorithms using a standard parameter set and random parameter sets.

The experimental results show that the proposed method offers two advantages over the other competing methods: the reliability in finding the optimal solution and the time required for finding the optimal solution.

This paper used sensitivity analyses as a tool to identify the suitable values for GA parameters for a groundwater monitoring problem. These values cause the algorithm to perform very well in terms of finding final optimal solution.

3. GA PROCESS

The basic GA process (Obitko, 1998) comprises the following steps:

(1) Initial population (solutions).
(2) Evaluate and calculate fitness function.
(3) Selection (top performance based on objective function).
(4) Crossover operation.
(5) Mutation operation.
(6) Elitism.
(7) Evaluation (go back to step 2): Obtain fitness values.
(8) End program based on number of generation.

3.1. Population Size

The population size should have a solution space that is large enough to be able to reach the optimal solution. If the population size is too small, the genetic algorithm may not explore enough of the solution space to consistently find good solutions. The initial population at each generation is improved toward the optimal solution.

3.2. Selection

Tournament method was used as selection process for this study. Parents are selected according to their fitness. This method compares two parents (strings) randomly based on the fitness value of the objective function. The one with the better fitness value (corresponding to optimal solution) wins and is selected for the initial population (mating pool). In this way, the parents (strings) with better fitness values have a higher chance of being represented more often than parents with poor fitness values, which have no chance at all to have a seat in the mating pool. In another word, better solutions have made themselves more than one copy in the new population and worse solutions have been eliminated from the population. Tournament selection has better convergence and computational time compared to any other reproduction operator (Goldberg and Deb, 1991). Selection alone cannot introduce any new children (individuals) into the population. The new children are generated by GA operators mainly by crossover and mutation.

3.3. Crossover

In this stage the initial population (mating pool) represents the population of parent strings that survive to create children strings. Unlike the selection stage, the crossover and mutation stages create new children or solutions. In the process of crossover, children strings are created with the aim of enhancing the results of the optimization.

Crossover emphasizes the exploitation of genetic material from the current population and assumes the genetic material necessary for an optimal solution currently exists in the population. The crossover operation is usually pairwise (two parents). The purpose of crossover is to add some convergence to the GA population, thus pushing the GA solution towards converging to a local optimum or global optimum.

The crossover is performed on each selected pair with a specified probability, referred to as crossover probability, which is typically between 0.7 and 0.9. The crossover can be performed using either the single-point method or the uniform method. The end result of crossover operation is the creation of a new population, which has the same size as the old population.

3.4. Mutation

Unlike crossover, the purpose of mutation is to add some diversity to the GA population, thus preventing the GA solution from prematurely converging to a local optimum. The mutation operates in a bitwise fashion. It operates on a single child or individual (the new child after crossover), producing a single new child. The basic operation of mutation is based on exploration with the purpose of generating a new individual whose representation potentially does not currently exist in the population. Mutation changes from 1 to 0 and vice versa according to the mutation probability.

The probability of any bit in a string being selected for mutation is controlled by the mutation probability. (Carroll, 1996) suggests the following formula:

$$p = 1/n \quad (1)$$

where p is the probability of mutation and n the number of variables.

As with crossover, a random number between zero and one is generated. If this number is smaller than the mutation probability, the mutation operation is performed, otherwise, it is skipped. It is carried out by flipping bits at random, with some small probability. The purpose of mutation is to maintain diversity within the population that alters one or more gene values in a chromosome from its initial state. This can result in entirely new gene values being added to the gene pool.

3.5. Elitism

When creating new population by crossover and mutation, there is a big chance that the best chromosome (child) will be lost. By using elitism, a small percentage of good and best chromosomes are carried into the new population in the next generation. The rest of the chromosomes are chosen by selection method. Elitism can very rapidly increase the performance of GA, because it prevents losing the best solution that has already been found and used to jump out of a local optimum.

3.6. Stopping Criterion

A new population is formed after the mutation. It is evaluated according to the fitness (objective) function. The next step is to check the stopping criterion. If a user-defined stopping criterion is met, or when the maximum number of generations allowed is reached the procedure stops, otherwise, the program goes back to the selection (reproduction) stage to perform another cycle (i.e., generation) of selection (reproduction), crossover, and mutation. As the GA population evolves over a certain number of generations, the optimal or a near-optimal solution is obtained. Usually, the objective function improves much more rapidly in earlier generations than in later generations.

4. Sensitivity Analysis

The ability of Genetic Algorithm (GA) methods, to identify optimal solutions in groundwater management optimization problems have been widely demonstrated (Abdeh-Kolahchi, 2006). However, one of the main concerns in applying these methods is identifying suitable values for the GA parameters, such as population size, number of generations, probabilities of crossover, probability of mutation, and the usage of elitism. The values selected for these parameters have a significant impact on the objective function solution as well as the time taken to find that solution. A considerable amount of time and effort must be dedicated to estimate these parameters for the GA to obtain optimal solution. The impact of each parameter will be dependent on the values of the other parameters. This paper describes sensitivity analyses to identify the suitable values for GA parameters for a groundwater monitoring network design problem. Through these analyses, suitable parameter values are identified, causing the algorithm to perform very well in terms of finding finial optimal solution.

Sensitivity analysis determines the model input parameters that have strong influence on calculated results. It plays an important role in parameter estimation and identification. This requires the evaluation of sensitivity with respect to every estimated parameter.

Sensitivity analysis results provide good knowledge of the importance of parameters as well as their best value ranges, by examining the overall response of GA results to changes in the input parameters. If the sensitivity analysis shows that a particular parameter is highly sensitive to the GA response, the value for the parameter should be carefully selected.

In sensitivity analysis the GA is established based on the best-estimated values for parameters. During simulation one parameter is varied by a certain percentage, while the other parameters are kept unchanged, to find the model response.

5. Sensitivity Analysis of Genetic Algorithm Parameters

To improve the GA performance, a series of sensitivity analyses are carried out on main parameters of GA, i.e., several runs with different probabilities of crossover and runs for the same code to justify the best parameters probabilities.

The sensitivity analysis is performed on most important parameters of genetic algorithms. Table 1 presents the parameters that are used in genetic algorithms to solve the problem of groundwater monitoring design.

Table 1. Genetic algorithm parameters

Parameters	Variable/Constant	Value
Penalty Parameter	Constant	1200
Initial Population	Constant	5000
Number of Generation	Variable	50000
Random Number	Variable	0.375
Crossover Probability	Variable	0.87
Mutation Probability	Variable	0.156
Elitism	Process	used

During the sensitivity analysis, values of one parameter were changed while the other parameters remained constant on the genetic algorithms code. A penalty function approach is required to satisfy the constraints of the problem, and experimentation is needed to determine proper settings of the penalty parameters. The value of penalty coefficient should be high enough, so that the GA could not violate the constraints.

6. Sensitivity Analysis of Random Number to Fitness

The results of sensitivity analysis of random number versus fitness value or objective function are presented in Table 2 and Figure 1. These results are dependent on the number of generations or stopping criteria, crossover probability, and mutation probability. The random number variable increases by increment 10 percent starting from zero up to one. The other parameters that remain constant are number of generation (50000), crossover probability (0.87) and mutation (0.0156). Based on the above condition the GA runs for at least 10 times. The results of this sensitivity analysis obtained based on the solution criteria are subdivided in three cases.

Table 2. Random number sensitivity analysis

Parameters	Condition	Value
Crossover Probability	Constant	0.87
Mutation Probability	Constant	0.0156
Maximum Generation	Constant	50000
Random Number	Fitness Value	Generation
0	-190.6933	2545
0.1	866.5391	1324
0.2	866.5391	1704
0.3	866.5391	43332
0.34	869.4457	9990
0.39	869.4457	3225
0.4	869.4457	20250
0.46	869.4457	28215
0.5	-8935.7610	24049
0.6	866.5391	1326
0.7	866.5391	1971
0.8	805.6685	577
0.87	869.4457	16430
0.9	866.5391	1866
1	-190.6933	2545

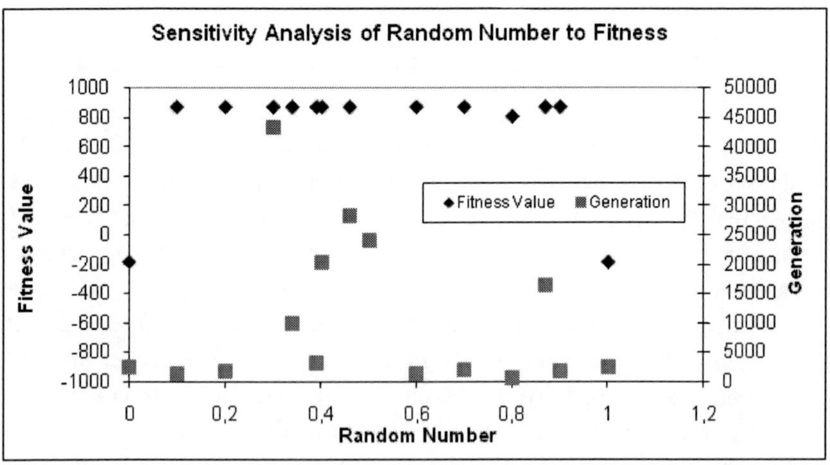

Figure 1. Random number sensitivity analysis.

6.1. Case One

When the values of the random number are 0, 0.5, and 1, the fitness value or objective function could not reach the exact solution (using above conditions) and the solution was trapped in local maximum.

6.2. Case Two

When random number values are 0.1, 0.2, 0.3, 0.6, 0.7, 0.8, and 0.9, the limitation of the stopping criteria, at 50000 generations, causes the objective function to not reach the exact solution. This may be overcome by increasing the stopping criteria to obtain the optimal solution.

6.3. Case Three

Other values of random numbers (0.34, 0.39, 0.4, 0.46, 0.8, and 0.87) fully satisfy the objective function and the exact solution is attained. The objective function reaches an exact optimal solution when the random number is equal to 0.39, which only takes 3225 generations.

7. Sensitivity Analysis of Crossover Probability to Objective Function

The results of sensitivity analysis of crossover probability versus objective function are presented in Table 3 and Figure 2. These results are under the constant conditions for the number of generations as stopping criteria, mutation probability, and random numbers, while the crossover probability variable is between zero and one with increment of 10%. The trend of the results is similar for almost any crossover probability value. While in most cases the objective function identifies the optimal solution with a fitness value of 869.4457, in two cases the objective function is trapped in a local maximum.

Table 3. Crossover probability sensitivity analysis

Random Number	Constant	0.375
Mutation probability	Constant	0.0156
Maximum Generation	Constant	32000
Crossover Probability	Objective Function	Generation
0.1	805.6685	615
0.2	805.6685	5198
0.3	869.4457	9788
0.4	869.4457	2344
0.5	869.4457	1662
0.6	869.4457	1532
0.7	866.5391	2573
0.8	869.4457	2154
0.9	869.4457	1285
1	869.4457	2777

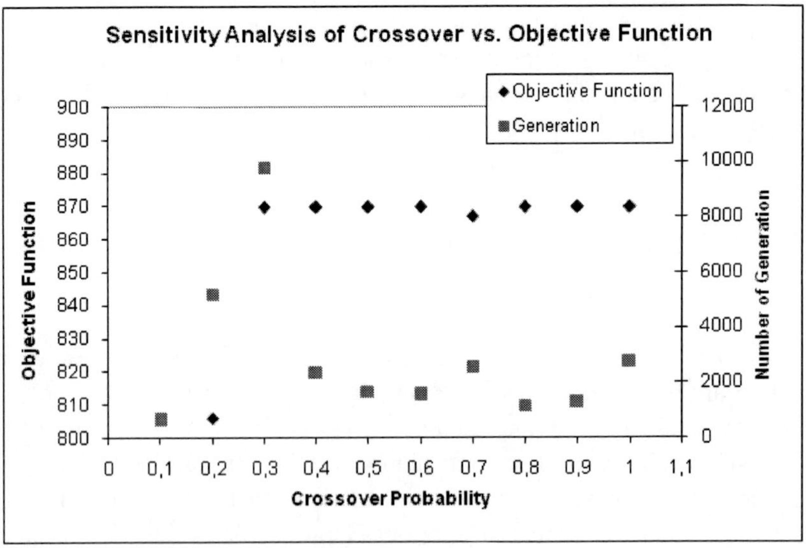

Figure 2. Crossover probability sensitivity analysis.

Although the crossover probability value greater than 0.2 reaches the optimal solution in this problem, as discussed earlier the crossover probability should be in the range of 0.8 – 0.90. In this case, fastest convergences (1154 generations) were obtained for a crossover probability value of 0.87.

It can be concluded from Figure 2 that the objective function is not very sensitive to the crossover probability for the groundwater monitoring network problem described here.

8. Sensitivity Analysis of Mutation Probability to Objective Function

The results of sensitivity analysis of mutation probability versus objective function are presented in Table 4 and Figure 3.

These results are under the constant conditions for the number of generations as stopping criteria, crossover probability, and random numbers, while the mutation probability varies between zero and one with increment up to 10%. Unlike crossover probability, mutation probability is highly sensitive to the objective function.

Table 4. Mutation probability sensitivity analysis

Crossover Probability	Constant	0.87
Random Number	Constant	0.375
Maximum Generation	Constant	32000
Mutation Probability	Objective Function	Generation
0	805.6685	683
0.01	805.6685	200
0.0156	869.4457	3679
0.02	805.6685	2342
0.03	869.4457	4423
0.04	866.5391	3410
0.05	869.4457	6157
0.06	869.4457	8091
0.1	799.6685	22064
0.2	-1136.833	4467
0.3	-2767.971	1360
1	866.5391	21026

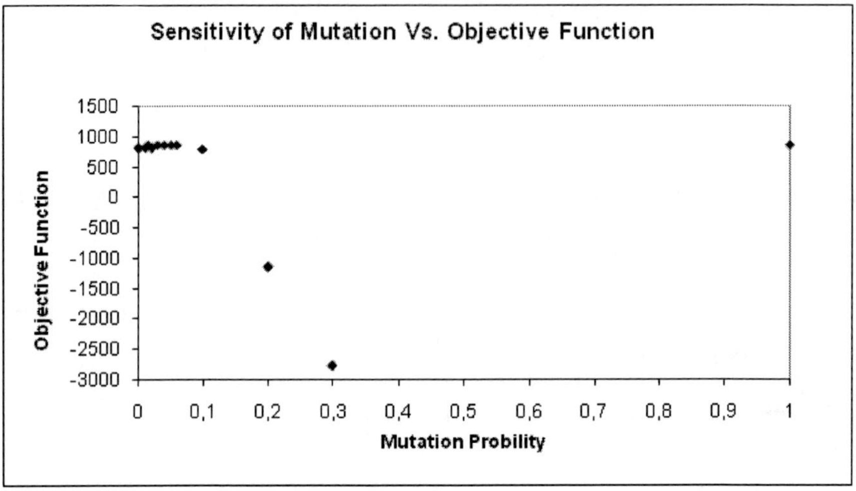

Figure 3. Mutation probability sensitivity analysis.

As suggested by Carroll (1996) the mutation probability is inversely related to the number of variables. For example, for the groundwater monitoring network design problem the total number of variables reached 64, so the mutation probability should be 1/64 or 0.0156, which is the value selected for this study. The results as shown in Figure 3 clearly indicated that the mutation probability with the presence of more than two mutations per chromosome leads to very poor GA performance. Mutation of one monitoring well out of the total number of 64 monitoring wells produces the best results. The sensitivity result shows that the mutation probability should be less than 0.1 for this case.

Increasing the mutation probability created a divergence from the optimal solution as Table 4 and Figure 3 show. The optimal solution converged at 0.87 crossover probability and 0.0156 mutation probability gets only 3679 generations.

9. Sensitivity Analysis of Objective Function with and without Elitism

To avoid elimination and loss of a good solution, which may occur through crossover and mutation, elitism guarantees that the good solution continues to the next generation. Without elitism the search for an optimal solution in this case not only takes time but also may not be achieved. By using elitism, we make sure that some small percentage of good and best solutions continues to the new population in the next generation.

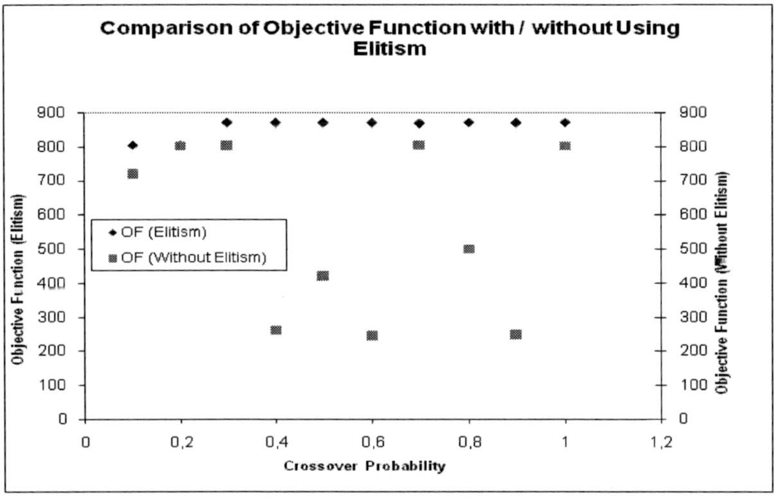

Figure 4. Elitism sensitivity analyses.

Table 5. Elitism Sensitivity Analysis

Random Number	Constant	0.375
Mutation probability	Constant	0.0156
Maximum Generation	Constant	32000
Crossover	Objective Function (Elitism)	Objective Function (Without Elitism)
0.1	805.6685	717.2128
0.2	805.6685	802.7134
0.3	869.4457	800.9769
0.4	869.4457	261.9967
0.5	869.4457	418.3011
0.6	869.4457	242.6637
0.7	866.5391	805.6685
0.8	869.4457	500.3737
0.9	869.4457	245.6107
1	869.4457	802.7184

Elitism can very rapidly improve performance of GA, because it prevents losing the current best solution. The sensitivity analysis results with and without elitism versus objective function are presented in Table 5 and Figure 4.

As seen from Table 5, the objective function achieves its optimal solution in most cases when elitism is included. Without the inclusion of elitism, optimal solution will not be achievable for the condition stated in the Table 5.

CONCLUSION

Sensitivity analysis determines the model input parameters that have strong influence on calculated results. Sensitivity analysis plays an important role in parameter estimation and identification. This requires the evaluation of sensitivity with respect to every estimated parameter.

To design an optimal monitoring network a combination of simulation of groundwater contamination and optimization tools such as genetic algorithms is necessary. Proper settings of GA's parameters are required to achieve the best performance. For this reason, a sensitivity analysis was carried out to determine the effect of GA's parameters especially crossover probability, mutation probability, and elitism on the GA performance. Similar analysis has been reported by Sharif and Wardlaw (2000) who found almost the same results for the effect of crossover and mutation probability.

From the analysis it can be concluded that the genetic algorithms are capable of obtaining optimal solution to the monitoring well design problem. Selecting correct parameters can speed up the search for the optimal solution. The sensitivity analysis of main features of GA, i.e., random number, crossover, mutation, and elitism, are discussed. The sensitivity analysis shows that for this particular problem, the value of crossover probability should be in the range of 0.80 to 0.90. The probability of mutation should be in the range of 0.01 – 0.10 which also is inversely dependent on the number of variables.

GA approach without elitism has some limitations. With very long chromosomes for groundwater monitoring network design problem, the likelihood of obtaining the optimum solution is reduced. Using elitism removes this kind of problem. Without the elitism, reaching the optimal solution for a large-scale problem such as a monitoring network is extremely difficult. The elitism not only guarantees for finding the best solution but also speeds up the search as well.

The performance of GAs has been found to have higher sensitivity to mutation probability and lower and stable sensitivity with respect to crossover probability. Therefore, proper choice of values for the parameters is essential for good GA performance. The sensitivity analysis has shown that very good results can be obtained for particular combination of crossover probability (0.87), mutation probability (0.0156), and using elitism. It can be concluded that the impact of mutation probability and elitism on the model performance is very significant and gives GA superior performance. It is shown that GA is efficient for solving groundwater monitoring network design problems. This approach significantly improves the results. Choosing the best values for GA parameters has a significant effect on the GA's operation.

ACKNOWLEDGMENT

The support from the Natural Sciences and Engineering Research Council of Canada (NSERC) Grant 43353 is gratefully acknowledged.

REFERENCES

Abdeh-Kolahchi, A. (2006) Optimal Dynamic Monitoring Network Design For Reliable Tracking of Contaminant Plumes in aquifer, *Civil Engineering*, Dalhousie University.

Carroll, D.L. (1996) 'Genetic algorithms and optimizing chemical oxygen-iodine lasers', In Developments in Theoretical and Applied Mechanics, Vol. XVIII, eds., H.B. Wilson, R.C. Batra, C.W. Bert, A.M.J. Davis, R.A. Schapery, D.S. Stewart, and F.F. Swinson, *School of Engineering*, The University of Alabama.

Davis, L., ed. (1991) Handbook of Genetic Algorithms, Nostrand Reinhold, New York, NY.

Deb K. and Agrawal, S. (1998) 'Understanding interactions among genetic algorithm parameters', In *Foundations of Genetic Algorithms*, 5, pp. 265–286.

Haupt, R. and Haupt, S. (1998) Practical genetic algorithms, John Wiley and sons.

Holland, J.H. (1975) *Adaption in Natural and Artificial Systems*, Univ. of Mich. Press, Ann Arbor.

Holland, J.H. (1992) *Adaptation in Natural and Artificial Systems*, 2nd Ed., MIT Press.

Obitko, M. (1998) Introduction to genetic algorithm, http://cs.felk.cvut.cz/~xobitko/ga/.

Sharif, M and Wardlaw, R. (2000) 'Multireservoir system optimization using genetic algorithms: case study', *J. Comp. in Civil Eng., ASCE*, 14(4), pp. 255-263.

Tongchim, S. and Chongstitvatana, P. (2002) 'Parallel genetic algorithm with parameter adaptation', *Information Processing Letters*, 82(1), pp. 47-54.

In: Sustainable Petroleum Operations
Editor: Rafiq Islam
ISBN 978-1-61324-157-8

Chapter 7

IMPACTS OF HYDROCARBON OPERATIONS IN THE WORLD'S LARGEST MANGROVE ECOSYSTEM OF BANGLADESH

M. Ibrahim Khan,[*1] ***Chefi Ketata***[2] ***and M. Rafiqul Islam***[2]

[1]Institute for Sustainable Energy, Environment and Economy, University of Calgary
ES-602 University Drive NW, Calgary, AB, Canada, T2N 1N4
[2]Faculty of Engineering, Dalhousie University-Sexton Campus
D510-1360 Barrington Street, Halifax, NS, B3J 2X4

ABSTRACT

Sundarbans mangrove ecosystem, the world largest mangrove forest in Bangladesh, is one of the most productive and biologically diverse wetlands in the world. This unique coastal tropical forest is among the most threatened habitats on earth. This paper critically examines the present status of the Sundarbans and assesses the potential threat of oil and gas exploration. Every aspect of the exploration and development of oil and gas is analyzed and the magnitude of impact quantified. To avoid or minimize the identified impacts a mitigation plan and guidelines are addressed. Since there is limited research done regarding these issues, this work can serve as a basis for understanding the effects of oil and gas in an environmentally sensitive area and point the way to further study.

[*] Corresponding Author Email: Ibrahim.Khan@Dal.ca.

Keywords: environmentally sensitive sites; Sundarbans; mangrove ecosystem; oil and gas; environmental effects.

1. INTRODUCTION

Mangrove ecosystems are one of the most productive and biodiverse wetlands on earth. However, these unique coastal tropical forests are among the most threatened habitats in the world (Fall, 2005). Due to commercial exploration they may be disappearing more quickly than the island tropical rainforest, and so far, with little public notice (Dewalt et al., 1996; Fall, 2005). The Sundarbans is the largest single-tract mangrove ecosystem in the world with an area of about 10,000 square kilometres (FAO, 1994). It contains a large variety of genera and species of plants, wildlife and astonishing biodiversity (Saha et al., 2006). It stretches across two countries Bangladesh and India, over the northern part of the Bay of Bengal (see Figures 1 and 2). Due to its great significance the Sundarbans was declared a World Heritage Site and Biosphere Reserve. However, the future of the Sundarbans remains uncertain due to logging, human settlement and other numerous exploitation activities (Saha et al., 2006).

In order to improve a struggling economy, the Bangladesh government has been looking into using the oil and gas resources in the country. After the recent discovery of huge amount of proven gas reserves in the country, Bangladesh got attention by many multinational oil companies. It is considered the future of the oil and gas is brighter and its reserves are placed at 15.3 trillion cubic feet (TCF), while the US Geological Survey estimates that Bangladesh contains additional 32.1 TCF "undiscovered reserves" (Kumaraswamy and Datta, 2006). The most terrestrial and offshore areas of Bangladesh have been divided into twenty three blocks and are gradually being leased to multinational oil and gas companies.

The Bangladeshi government recently completed the initial signing of a production sharing contract with two multinational oil giants to begin oil and gas exploration in block number five within the Sundarbans. The companies plan to conduct seismic and aerial surveys. In Bangladesh the gas and oil exploration business is in its infancy and therefore environmental impacts have not yet been felt. However, the impact of hydrocarbon pollution has been clearly reported in many other parts of the world (Stejskal, 2000; Teresa et al, 2002; Cheevaporn and Menasveta, 2003; Guzzella et al., 2005). Oil and gas exploration might pose a new and, for the most part, unknown threat to the Sundarbans ecosystem.

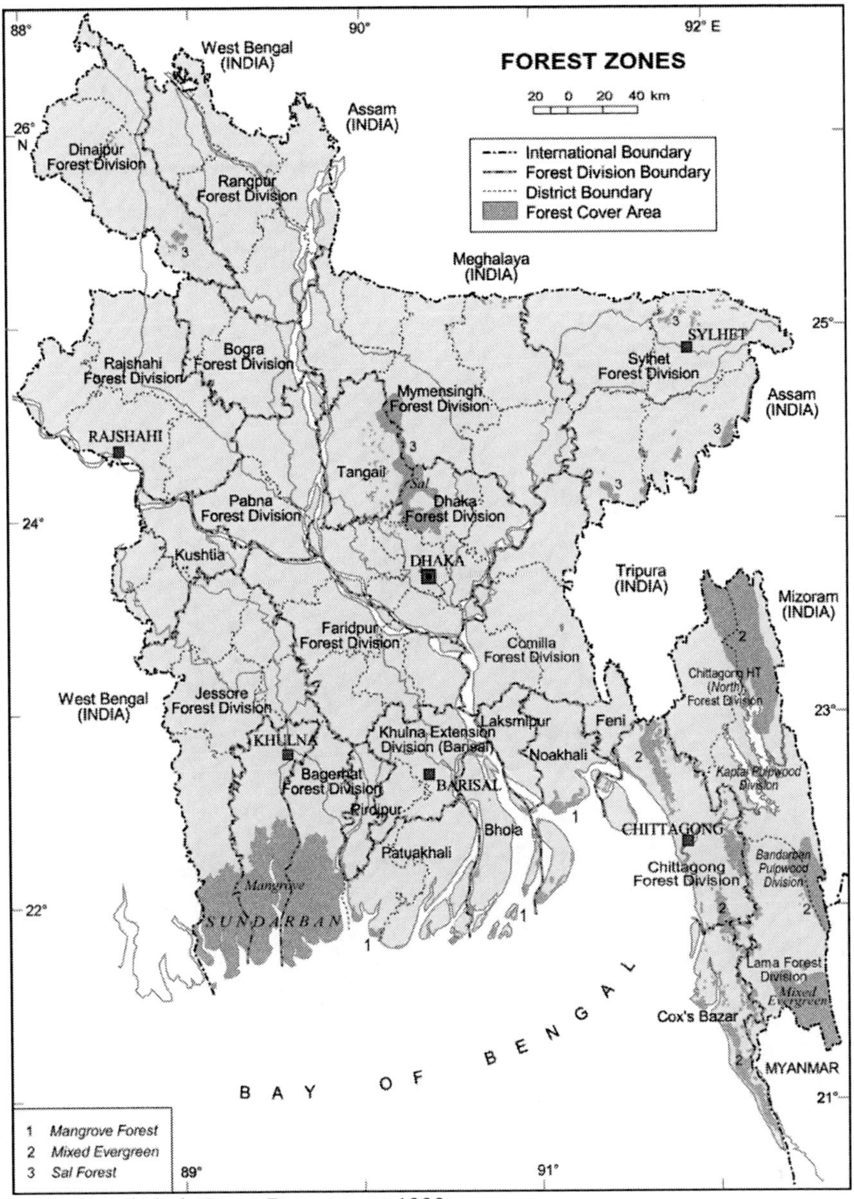

Source: Bangladesh Forest Department, 1999.

Figure 1. Forest zones map of Bangladesh.

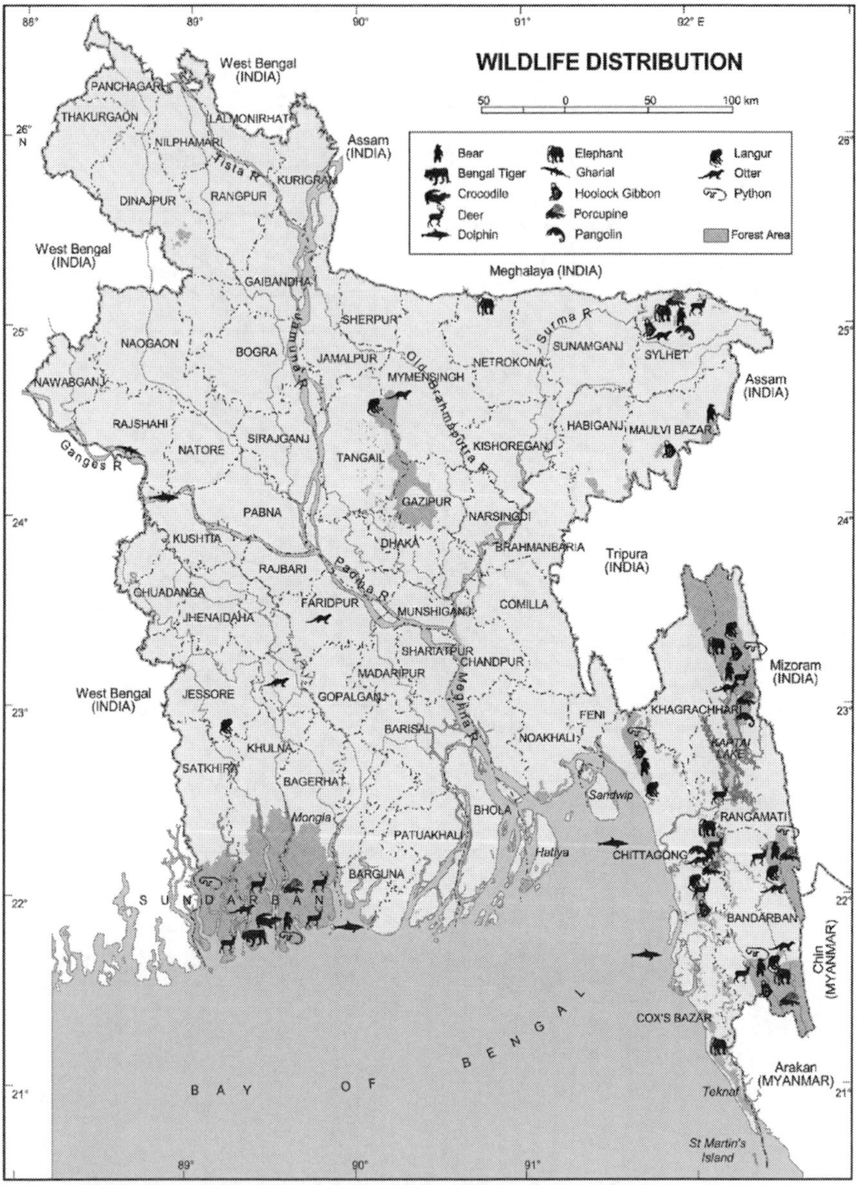

Source: Government of the People's Republic of Bangladesh, 2006.

Figure 2. Wildlife distribution map of Bangladesh.

There has been a very limited amount of research done so far to determine the effects of oil and gas exploration in mangroves ecosystems except for the preliminary work of Stejskal (2000) and Teresa et al. (2002). The main objective of this paper is to critically examine the potential threat of oil and gas exploration and production on the Sundarbans mangrove ecosystem and provide guidelines to avoid or minimize effects. This paper initially provides a brief description of the present resources status of Sundarbans, followed by detailed analyses of different oil and gas exploration and development activities. Then the impacts of these activities are quantified. Finally, different measures, mitigation plan, and guidelines are discussed.

2. Environmental Engineering Planning and Impact Analysis

The environmental engineering planning and impact analysis for oil and gas activities in a certain area should consider its physical, social, economic, and environmental impacts as a whole in order to foster a healthy and efficient society in the investigated region. Environmental engineering planning is a continuous process. It is composed of the following stages (see Figure 3):

1. Goals statement.
2. Data collection and analysis.
3. Program establishment.
4. Resources allocation.
5. Actions implementation.
6. Evaluation.

The environmental impact assessment of an oil and gas activity comprises the following steps (see Figure 4):

1. Site location.
2. Activity type.
3. Impacts identification.
4. Proposal.
5. Mitigation and impact management.
6. Evaluation.

Impact assessment is based on oil and gas activity nature and substances released while performing such activity.

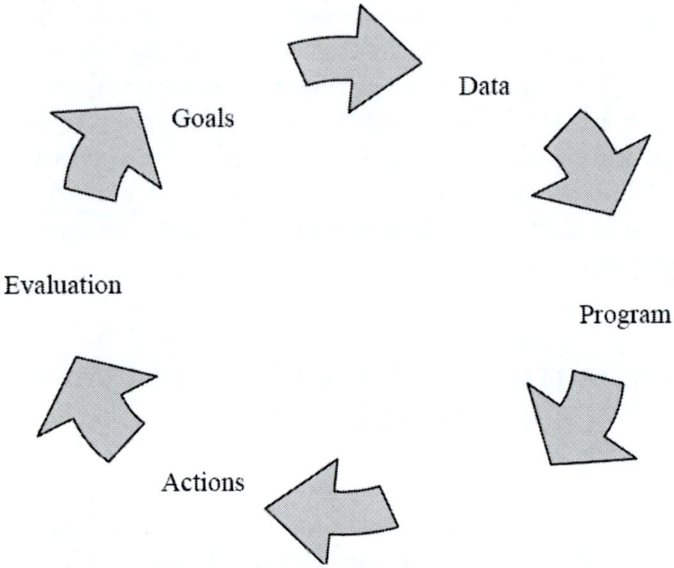

Figure 3. Environmental engineering planning stages.

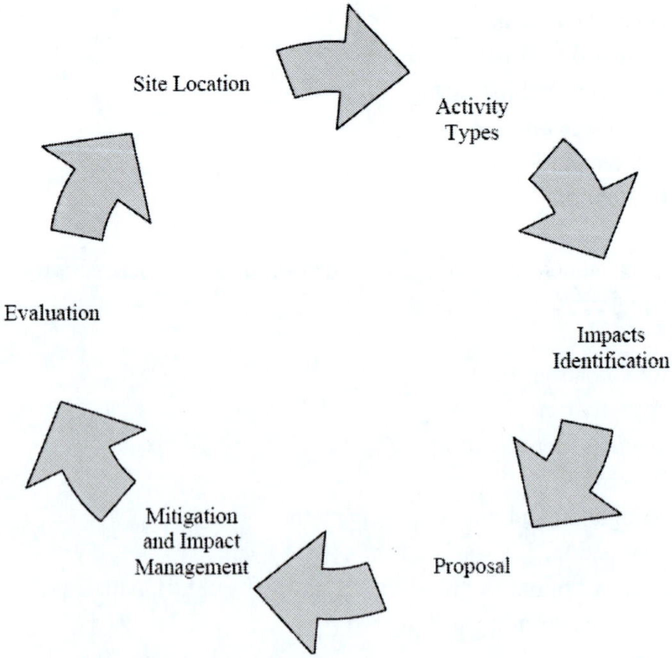

Figure 4. Environmental impact analysis steps.

Table 1. The population of the large terrestrials mammals of Sundarbans

Name	Population
Spotted Deer	80,000
Wild Boar	20,000
Tiger	350
Rhesus Macaque	40,000
Otter	20,000

Source: ESCAP, 1987.

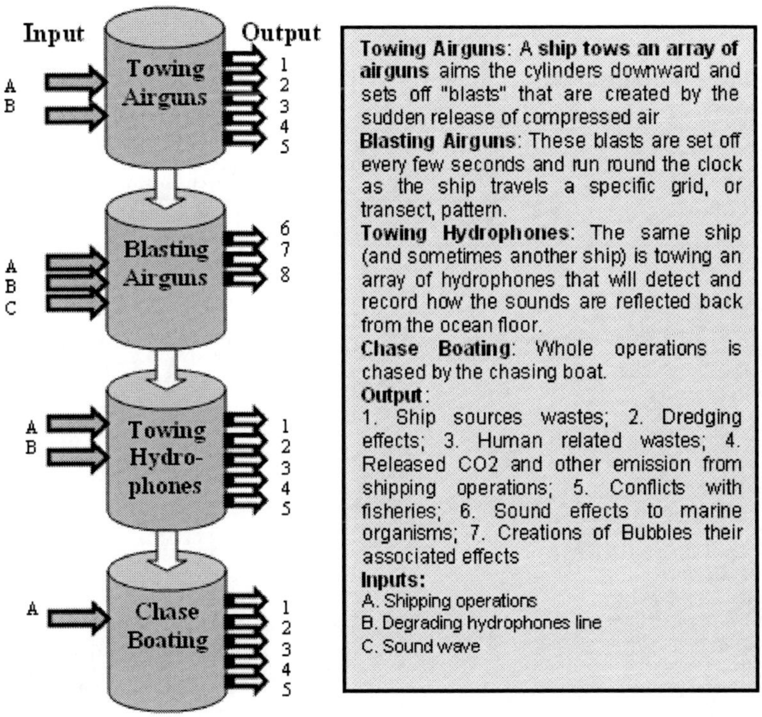

Figure 5. Schematic Seismic Activities in offshore exploration involving towing airguns, blasting airguns, towing hydrophones and chase boat operations (source: Khan and Islam, 2006a).

3. Sundarbans

The world's largest mangrove ecosystem, 'Sundarbans' locally means beautiful forests (see Figures 1 and 2). It is a dynamic, fragile, complex, and productive ecosystem in delicate balance with the factors of soil, water, and environment. It has the characteristics of marine and terrestrial, water and land, forest and sea, and a unique habitat of thousands of species, such as fish, shrimp, birds, royal Bengal tiger, crocodile, other wildlife species and numerous flora and fauna (Saenger and Siddiqi, 1993; Hossain, 2001; Islam, 2003; Saha et al., 2006). The major wildlife species are shown in Figure 5 and Table 2. This habitat has a unique position, not only in forestry but also in the culture and heritage of Bangladesh (Hossain, 2001; Islam, 2003). As a forest, Sundarbans constitutes about 44% of the forest area of the country and generates major revenue.

Table 2. Potential environmental disruptions resulting form oil and gas and development activities in the Sundarbans

Activities	Potential Development					
	Noise	Aircraft	Human intrusion	Traffic & access	Struct-ure	Habitat Altert	Harmful material
Land Survey		+	+	+			
Seismic trail clearing	++		+	+		+	
Seismic wave production	+++	+	+				
Clearings Vegetation	++		+			+	
Road Construction	+		++	+		+	
Mobilization of truck/equip.	+	+	+	+			
Site Development	++	+	++	+	+++	++	+
Drill pad construction	+		+			+	
Excavation of storage/mud pits	+						
Drilling and related activities	+		+		+		
Water supply	+				+		+
Borrow pit excavation	+		+				
Wellhead/pump until installation							+
Construction of process plant & storage facilities	+		+++	+	++	++	+
Installation of flow lines			+			+	
Erection of power lines	+		+		+		
Communication system					+		
Drainage facilities			+		++	+	
Operation of process facilities	+					+	
Pipe string	+						
Trenching and pipe installation	+		+		+		
Pipe burial and backfill	+		+				
Maintenance and inspection	+		+				
Accidents	+					++	+
Secondary recovery		+	+			+	
Air trip/helipads		+					
Worker Accommodations	+		+				
Increase in local population	+		+++		+++	+++	
Development of ancillary - Industry	+		+	++	+++	+++	
Well plugging			+				
Site restoration/renegotiation			+				

4. Oil and Gas Exploration and Development Activities

4.1. Seismic Exploration

Seismic study provides primary information about geological formation of a study area (Khan and Islam, 2006a). Using airguns, sonar waves are transmitted to subsurface and these waves travel back and are received by hydrophones. Technological details of seismic operations for the offshore oil and gas operations are reported by Khan and Islam (2006a). There are different types of technologies used for seismic study, such as 2D, 3D and 4D (Khan and Islam, 2006b). The Sundarbans consist of land and water. As a result, offshore and onshore seismic operations are needed. Considering the environmental condition of Sundarbans potential exploratory activities are shown in Table 2.

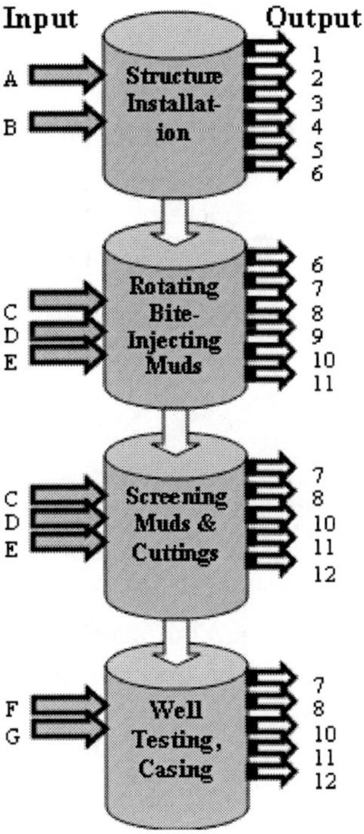

Structure Installation: Different types of rigs are selected for hosting, performing circulating and rotation system, backing-up pressure-control equipment and holding many pumps.
Rotating Drill Bite and Injecting Muds: Motorized equipment rotates the drill pipe to make it cut into rocks, and prime movers circulate drilling fluids.
Screening Muds and Cuttings: The muds flow out of the annulus above the blowout preventer over the shale shaker screen to remove formation cutting, and back into the mud tanks
Well Testing and Casing: When the drill pipe goes deeper it needs casing, generally after 1000m.
Output:
1. Ship source wastes; 2. Dredging effects; 3. Human related wastes; 4. Releases of CO2; 5. Conflicting with fisheries; 6.Shipping effects; 7. Drilling muds; 8.Drilling cuttings; 9. Flare; 10. Radio-active materials; 11.Produced water; 12.Release of injected chemical
Input:
A. Associated inputs related to Installation of huge structures
B. Use of supply ships
C. Water-based drilling muds
D. Oil-based drilling muds
E. Synthetic-based drilling muds
F. Well testing fluids
G. Inputting casing

Figure 6. Systematic diagram of drilling activities and associated waste releases (source: Khan and Islam, 2006b).

4.2. Drilling

Once the seismic study indicates the possible occurrence of oil, drilling of an exploratory well may be initiated to determine if oil and gas of commercial quantity and quality exist (Khan and Islam, 2003 and 2006; Khan et al, 2006). The actual drilling operation normally uses a rotary drill bit and drill string, consisting of the drillpipe, the bottomhole assembly, and other tools used to make the drill bit rotate and gradually bore into the earth. Drilling is considered the most destructive phase among total oil and gas operations (Stejskal, 2000; Khan and Islam, 2003; Khan and Islam, 2006a). The drilling activities and their waste generation are shown in Figure 6 and Table 2.

4.3. Development and Production

Oil and gas field production begins if a commercially feasible amount of oil and gas is discovered in drilling. It starts by constructing many production and transportation facilities and replacing temporary facilities. Based on the size of the oil reservoir, immense amounts of materials are necessary to erect production facilities. For example, the Sable Island offshore energy project in Canada used more than 900 kilometers of pipe, 109,000 tonnes of steel, 53 kilometers of chrome tubing, and 160,000 tonnes of concrete (Khan et al, 2006). The potential development and production activities for the Sundarbans mangroves are shown in Table 2.

4.4. Decommissioning

Decommissioning is the last step of oil and gas operations. However, in every phase some decommissioning activities take place (dry holes and depleted producing wells are plugged with cement). Drilling rigs and support equipment are removed from an unsuccessful well (Khan and Islam, 2006a). When the entire lease is completed all structures must be decommissioned. The activities related to the Sundarbans project are shown in Table 1.

5. VALUED ECOSYSTEM COMPONENTS OF SUNDARBANS AND POTENTIAL IMPACTS

Many studies report that mangrove ecosystems especially are unique among the most productive ecosystem. To identify the potential impact of oil and gas development, the valued ecosystem components (VEC) of Sundarbans are identified. Considering the values and services provided by Sundarbans five VECs are selected which are discussed in this section.

5.1. Wildlife

The Sundarbans is a unique habitat for wildlife and it supports a very rich and diverse fauna (Saha et al., 2006). There are 315 species of birds, 400 species of fish, 53 species of reptiles, and 8 species of amphibians in Sundarbans (Hendrichs, 1975; Mukherjee, 1975; Sarker and Sarker, 1988; Islam, 2001). Species diversity and wildlife population are shown in Tables 1 and 3.

Oil and gas operations can cause numerous impacts on the wildlife of Sundarbans. Oil and gas development causes problems (Bolze and Lee, 1989). For example, the presence of development-associated structures may interfere with movement (see Table 4). If it is severe or prolonged, it may result in reduced reproductive success or loss of available habitat (Bolze and Lee, 1989). Increased human/wildlife encounters may cause animals to avoid an area. It is also reported that effects on courtship behaviour reduce existing stock and permanent loss of habitat use (Bolze and Lee, 1989). The noise of helicopters and other crafts will be a serious threat to the huge winter migration of birds. A detailed impact on wildlife is shown in the Table 4.

5.2. Species at Risks

There is an amazing number of wildlife throughout the forest, but a good number of valuable animals have become extinct (Sarker and Sarker, 1988). For example, out of the 49 mammal species no less than four major species, the Javan rhinoceros (*rhinoceros sonadacius*), wild buffalo (*Buhalus huyabails*), swamp deer (*Cervus duvauecli*) and hog deer (*Axis porcinus*) have become extinct since the beginning of the century (Sarker and Sarker, 1988).

Table 3. Species diversity in the Sundarbans

Group of Species	No of Species
Mangroves	27
Mangrove Associate Plants	69
Phytoplankton and Algae	37
Ichthyoplankton	200
Fish and shrimp	400
Shark	3
Shrimp	24
Skates and Rays	5
Crab	7
Lobster	3
Molluscs	26
Mammals	49
Reptiles	53
Amphibians	8
Birds	310

Sources: Hendrichs, 1975; Mukherjee, 1975; Ahamed, 1984; Islam, 2001.

The tiger is one of the most endangered species in the world. Today, there are only about 7,000 tigers left in the wild, down from more than 100,000 a century ago (UN, 2002). Almost 550-600 tigers reside in the Sundarbans. The following Sundarbans species are those considered to be globally threatened as of late 1993 according to the IUCN status categories: White-bellied Heron *(Ardea insignis*), Lesser Florican (*Eupodotis indica*), *Panthera tigris* (Tiger), Hispid Hare (*Caprolagus hispidus*), Hoolock Gibbon (*Hylobates hoolock*), Green Turtle *(Chelonia mydas*), Hawksbill Turtle (*Eretmochelys imbricata*), Olive Ridley (*Lepidochelys olivacea*), Batagur (*Batagur baska*).

Oil and gas developments can further endanger the species which are already in risk. Some studies from different parts of the world show that oil and gas development activities have negative impacts on species at risk (Stejskal, 2000). Table 4 shows the detailed impacts of oil and gas operations on the Sundarban wildlife. Similar effects are also discussed in the wildlife section.

5.3. Special Habitat

Sundarbans provides a critical habitat for diverse marine and terrestrial flora and fauna (Table 3). It provides ecological services and works as an essential habitat, nutrient producer, water purifier, nutrient and sediment trap, storm barrier, shore stabilizer, aesthetic attraction and energy storage unit (Hossain, 2001; Islam, 2003). Mangrove sediments appear to be efficient in retention and accumulation of phosphorus (Silva and Mozeto, 1977).

Table 4. Potential impacts of oil and gas development activities on the Sundarbans' VECs

Activities	Potential Impact Levels on VECs						
	WILD-LIFE	*SPECIES AT RISK*	*ESPECIAL HABITAT*	Forest	Fisheries	Aqua-culture	Solar salt
Land Survey	2	2	2	1	0	1	1
Aerial Survey	2	2	2	0	0	2	0
Seismic trail clearing	2	3	3	5	3	3	0
Seismic wave production	7	7	2	1	5	7	0
Clearings Vegetation	5	7	7	6	6	3	0
Road Construction	4	5	6	3	3	1	1
Mobilization of truck/equip.	3	4	3	1	1	1	1
Site Development	5	5	5	3	3	0	1
Drill pad construction	4	4	5	3	2	1	0
Excavation of storage/mud pits	4	3	4	3	1	1	1
Drilling and related activities	7	7	6	2	8	6	7
Water supply	0	2	4	5	1	3	2
Borrow pit excavation	2	2	2	1	1	1	2
Wellhead/pump until installation	2	2	2	1	1	0	0
Construction of process plant & storage facilities	3	3	3	2	4	0	0
Installation of flow lines	4	3	3	2	2	1	0
Erection of power lines	2	3	4	3	0	0	0

Table 4. Continued

Communication system	1	2	1	2	0	0	0
Drainage facilities	5	2	5	3	3	2	7
Operation of process facilities	6	3	4	2	5	3	7
Pipe string	4	3	3	2	2	2	1
Trenching and pipe installation	2	4	3	6	1	3	2
Pipe burial and backfill	2	3	2	3	1	2	2
Maintenance and inspection	2	4	1	1	2	1	2
Accidents	7	10	10	7	8	9	10
Secondary recovery	5	5	7	5	3	3	4
Airstrip/helipads	3	3	4	2	1	2	0
Worker Accommodations	4	4	4	1	1	1	0
Increase in local population	5	5	7	3	3	1	0
Development of ancillary industry	5	6	5	3	2	1	1
Well plugging	2	3	2	1	2	1	2
Site restoration/ renegotiation	2	4	2	2	2	2	3

Mangroves have a significant importance for fisheries. Many species use the mangrove ecosystem in various ways (Hossain, 2001; Islam, 2003). Some fishes, obligatory species, spend their entire lifecycle in this environment. Migratory species spend a crucial part of their life (e.g., peneaid prawns) using it as a shelter or as a source of food. Others, facultative, survive and reproduce even in the absence of mangroves but show a preference for the habitat and the nutrient provided therein (Hong, 1996).

Healthy mangrove forests are key to healthy marine ecology. Deterioration of the Sundarbans Mangrove due to oil and gas exploration might directly and indirectly cause declines of fisheries, degradation of clean water supplies, salinization of coastal soils, erosion, and land subsidence (Hossain, 2001; Islam, 2003). This coastal ecosystem is so specialized that any minor variation in its hydrological quality causes noticeable mortality (Blasco et al., 1996). Potential impacts of oil and gas exploration and development are presented in Table 4.

5.4. Mangrove Forests

The Sundarbans comprises 45% of the productive forest of the country. It contributes half of the forest-related revenue and is an important source of wood and no-wood resources (Katebi, and Habib, 1989). A wide range of forest products such as timber, fuelwood, pulpwood, matchwood and thatching materials are obtained from the Sundarbans. In 1992-93, US$ 7.0 million revenue was earned from the Sundarbans (FAO, 1994).

The natural vegetation of the Sundarbans is composed of halophytic tree species commonly termed as mangrove. There are 56 species of mangroves and mangrove associates in the Sundarbans (see Table 3) (Chaffey and Sandom, 1985; Khatun and Alam, 1987; Saenger and Siddiqi, 1993). The major predominant tree species are Sundaries (*Heritiera fomes*) and gewa (*Exocoecaria agallecha*) followed by minor proportions of passur (*Caraoa obovata*), keora (*Sonneratia apetala*), baen (*Avicennia officinalis*), kankar (*Pruguiera gymorhiza*), and a few other species like goran (*Ceriaps rexburghias*), hantal (*Pheoenix paludosa*), shingra (*Cynometra remiflora*), khalisi (*Ameiceras maius*), bhola (*Hibiscus tiliaceus*). Oil can have a significant impact on the mangroves. Potential impacts on mangrove forest are shown in the Table 4. If an oil slick were to enter the mangrove forest when the tide is high, it would be deposited on the aerial roots and sediment surfaces as the tide proceeds. Oil that covers the trees' breathing pores can asphyxiate the surface roots and lead to the death of the mangrove (Katebi, and Habib, 1989). The trees can also be killed by the toxicity of the hydrocarbon. It is reported that a lighter fraction of the oil, considered to be the most toxic, generally evaporates or degrades rapidly and the heavier fraction is the cause of most of the chronic impacts (Teal and Howarth, 1984). Chronic exposure to hydrocarbons results in damage to aerial roots, 25-64.5% defoliation depending on the species, 100% mortality in seedling, reduction in propagate density accompanied by atrophy and malformations, and finally a serious reduction in basal area (Lamparelli et al., 1997). Recovery from the effects of oil spills begins 10 years after an incident. It is reported that an oil spill in 1994 near the Sundarbans spread about 15 km downstream from the ship and caused the instant mortality of seedling of *Heritiera*, *Exoecaria*, and many species of grasses (Karim, 1994).

5.5. Coastal Fisheries

In the Sundarbans mangroves, an average of ten thousand tons of fish and shellfish was caught annually in the late seventies and early eighties (KU, 1995). With 169908 km^2 of waterways, approximately 53 kg/ha of fish are being caught within the mangrove area itself (Penn, 1983; Mahmood et al., 1994; KU, 1995). To this figure should be added the portion of the offshore fisheries of mangrove development species. Whereas over 120 species of fish and shellfish are known to be caught of mangrove-dependent species in the Sundarbans area (see Table 3), the main portion consists of shrimps and hilsha (*Tenolosa* and *Hilsha* sp.).

A recent estimate of the total yield from the inshore-estuaries and offshore fishery of the Sundarbans reserved forest records 11700 tons in the year 1993 (KU, 1995). Penn (1983) estimated the potential annual production of demersal finfish in the offshore waters of Bangladesh to be 10000 to 20000 tons, and that of the shrimps to be 200 to 4000 tons. He mentioned that available yields per unit length of the coastline were relatively high by world standards. One of the main reasons for this high yield is the influence of the Sundarbans.

Potential impacts of oil and gas operations are shown in Table 4. Some of them are:

- Reproduction of fish species;
- Damage to fish eggs and larvae;
- Reduced fish stock;
- Change in the quality of fish and shrimp due to tenting;
- Effects on the fish migration;
- Depletion of nutrient form the fish nursing ground through reducing litter fall.

5.6. Coastal Aquaculture

Coastal shrimp farming is one of the major economic activities in Bangladesh and shrimp ponds are generally developed in the mangrove areas. Chakaria Sundarbans, the southeast mangrove areas of Bangladesh, are almost (90%) destroyed for shrimp farming (Mahmood et al., 1994) (see Figure 6). Shrimp farming is not permitted in the southwest Sundarbans reserve forests but recently several aquaculture farms have been established at the border of the forest (FAO, 1984a). Frozen food export is the second largest sector in Bangladesh. The major portions of the earnings come from farmed shrimp export. Every year around 30000 tons of tiger shrimp is exported to Japan, North America and Europe.

Exploration and development of oil and gas may cause environmental problems, such as tainted product quality, increased disease rate due to the oil pollution, closure of aquaculture production if pond soil is affected, and transfer of hydrocarbon to the human body. Details impacts are shown in Table 4. In the table, predefined schematic scale is followed. The degree of importance is within the range of 1 to 10, where 10 highest score and 1 lowest.

5.7. Salt Production

Solar salt production is a traditional and important industry in Bangladesh and produces edible salt. Traditionally, salt is produced in the muddy coastal land (where water retention power is high) by evaporating the seawater (salinity 35 ppt). In 1997 from December 12 to the end of March about 500,000 tonnes of refined salt were already produced in the coastal areas of Bangladesh (AMITECH, 1997). A total of 50,000 acres has been brought under salt production in the coastal areas (AMITECH, 1997).

Oil and gas operations in the Sundarbans as well as surrounding coastal areas can impact salt production causing quality deterioration of edible salt, hydrocarbon transfer to human body, salt bed contamination, closing of production, and pressure on the foreign currency (US dollar) to import salt. More detailed impacts are shown in Table 4.

6. Special Measures for Minimizing Effects on Sundarbans

From the above discussion, it is revealed that oil and gas activities have different impacts on the valued ecosystem of Sundarbans. Figure 7 shows the total score of each valued ecosystem components. Fisheries and especial habitat can be highly affected, because they got highest scores. Solar salt got lowest scores, which represents minimal effects. In addition, the mangrove ecosystem is different from any other resource. It is a very complex and dynamic ecosystem (Hossain, 2001; Islam, 2003). It is composed of multiple resources. Therefore, the Sundarbans need special kinds of measures to minimize the effects of oil and gas development. Guidelines for spatial, temporal and operational management are discussed in the following sections.

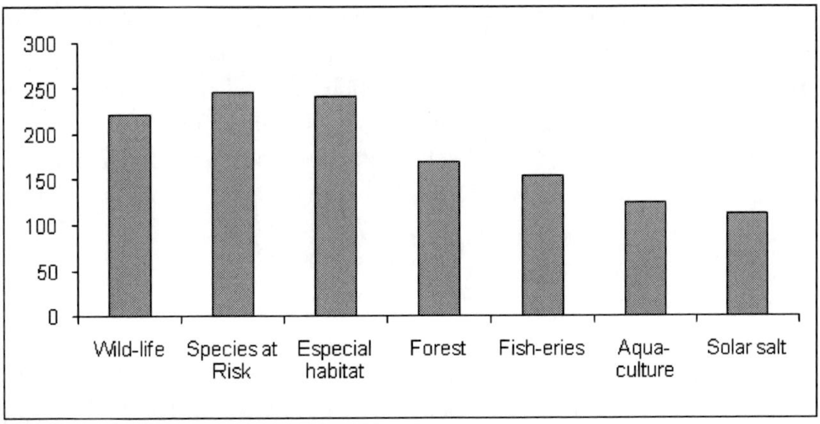

Figure 7. Impacts levels of oil and gas development on the VECs.

6.1. Spatial Management

- Infrastructure development must be avoided in the places that are sensitive and critical to wildlife, fisheries, flora and fauna, and other resources. In particular, highly detrimental activities, which are shown in Tables 1 and 2, must be avoided.
- Some of the activities are roads, facilities, dikes, and structures.
- The critical places that should be taken into account are wildlife breeding and nursery areas, fish breading ground, habitat of important species such as crocodiles and dolphins, and critical habitat for the species at risks.
- Sufficient and much wider buffer zones can be created to reduce wildlife visual contact with roads and development activities. Buffers may include topographic barriers, vegetation, and/or distance. Fences can also be used as a barrier to control the human movement in the ecologically critical areas.
- Provision for security areas, especially for ungulates, which contain necessary habitat elements and are shelter from disturbance. This may be accomplished by restricting road activities to one out of two

adjacent drainages at any one time. Ridgelines should not be developed in any way.

- Considering the impacts of aircraft activities (see Tables 1 and 2) the use of helicopters should be restricted. There should be restriction of aircraft activity to defined flight paths which avoid sensitive areas.

6.2. Temporal Management

- Sundarbans is multi-resourced ecosystems and different species have different breeding seasons. Restriction of activities to seasons and times of noncritical wildlife, fisheries, aquaculture, and other resources use.
- Oil and gas companies should compile a list of dates when the development activities should be prohibited in certain locations due to specific needs, such as breeding, migration or local movements, etc.

6.3. Operational Management

- Avoid making any dikes or embanking in the important habitats, such as crocodile grounds, dolphin channels.
- Avoid road construction, which will minimize clearing and cutting of mangrove vegetation.
- Worker shade should be constructed outside of the mangroves. Temporary water barge houses can be used.
- Waste disposal should be controlled by reusing or minimizing the wastes.
- Oil spill significantly damage the mangrove vegetations, because their respiratory root system is clogged by the oil. There should be a proper contingency plan for the oil spills.
- Seaward mangrove should be maintained as buffer to prevent erosion and cyclone intensity.
- Sundarbans has huge networks of rivers and channels. Using the riverian facilities the aircraft uses and piled roads construction can be avoided.
- Most of the Sundarbans soil is acid sulfated. It will need proper spoil disposal. Dredging should be avoided in the acid sulfate soil.

- Tidal and freshwater flow is vital for the mangrove vegetation. Therefore, water flow off should not be blocked through road construction and other development activities.

Conclusions

In this study, Sundarbans mangroves ecosystem is identified as a unique, productive, diverse, and precious wildlife habitat, in particular the rare tiger habitat, and areas suitable for estuarine crocodiles, and fresh water dolphins. It has huge ecological, economic and societal benefits. Historical uses of Sundarbans, such as shrimp farming, created many problems. Oil and gas operations are being considered as a new economic activity.

To identify the potential impacts of oil and gas operations different development and technological phases, such as seismic exploration, drilling, production and development, and decommissioning, are analyzed. The magnitude of impacts is quantified with the respect of different valued ecosystem components, such as wildlife, fisheries, and mangroves. This work is not intended for use in deciding whether or not to allow oil and gas development in Sundarbans but rather to aid in the identification of problems, to increase the manager's awareness of the implications of development, and to provide information that minimizes harmful effects. Multidisciplinary and long-term research work is needed, which will take into account the concerns detailed in this paper regarding the impacts of oil and gas operations on mangrove forests, fisheries, wildlife, and other resources of the Sundarbans.

Acknowledgment

The authors would like to thank the Atlantic Canada Opportunities Agency (ACOA) for funding this project under the Atlantic Innovation Fund (AIF).

References

Ahamed, M.K. (1984). Behundi net fishery in the brackishwater area of Patuakhali with emphasis on bionomics and mortality of shrimps.

Research Report Freshwater, Fisheries Research Station, Chandpur, Bangladesh (8): 36 pp.

AMETECH.. AMITECH (Academic of Technology), 1997. Salt production. A daily internet edition of Bangladesh news. April 12, 1997.

Blasco, F; P. Saenger, and E. Janodet. (1996). "Mangroves as Indicators of Coastal Change". *CATENA*. Vol. 27 (3-4): 167-178.

Bolze, D.A. and Lee, M.B. (1989). Offshore oil and gas development: Implications for wildlife in Alaska, *Marine Policy*: Volume 13, Issue 3: 231-248.

Chaffey, D.R. and J.H. Sandom (1985). Sundarban Forest Inventory Project, Bangladesh: A glossary of vernacular plant names and a field key to the trees. Overseas Development Authority (UK), Land Resources Development Center, Surrey, 23pp.

Cheevaporn, V. and Menasveta, P. (2003). Water pollution and habitat degradation in the Gulf of Thailand, *Marine Pollution Bulletin*, Volume 47, Issues 1-6: 43-51.

Dewalt, B.R., Vergne, P., and Hardin, M. (1996). Shrimp aquaculture development and the environment: People, mangroves and fisheries on the Gulf of Fonseca, Honduras, *World Development*, Volume 24, Issue 7: 1193-1208.

ESCAP (Economic and Social Commission for Asia and the Pacific), 1987. *Coastal environmental management plan for Bangladesh*, Vol. II-Main Report, Bangkok, 149 pp.

Fall, P.L. (2005). Vegetation change in the coastal-lowland rainforest at Avai'o'vuna Swamp, Vava'u, Kingdom of Tonga, *Quaternary Research*, Vol. 64, No. 3: 451-459.

FAO. (1984). Integrated development of Sundarbans, Bangladesh. Fisheries integrated development in the Sundarbans, based on the work of H.R. Rabanal. FAO: TCP/BGD/2309. 75 pp.

FAO. (1994). Mangrove forest guidelines. FAO Forest. Pap. (117).319 pp.

Government of the People's Republic of Bangladesh (2006). *http://www.bangladeshgov.org*.

Guzzella, L., Roscioli, C., Viganò, L., Saha, M., Sarkar, S.K., and Bhattacharya, A. (2005). Evaluation of the concentration of HCH, DDT, HCB, PCB and PAH in the sediments along the lower stretch of Hugli estuary, West Bengal, northeast India, *Environment International*, Vol. 31, No. 4: 523-534.

Hendrichs, H. (1975). The status of the tiger (Panthera tigiris) (Linne) in the Sundarbans mangrove forest (Bay of Bengal). *Saugetier-kundliche Mitteilugen*. 23: 161-199.

Hong, N.P. (1996). Mangrove destruction for shrimp rearing in Minh Hai, Vietnam: its damage to natural resources and environment. *SEAFDEC Asian Aquaculture* 18: 7-11.

Hossain, M.S. (2001). Biological aspects of the coastal and marine environment of Bangladesh, *Ocean and Coastal Management*, Volume 44, Issues 3-4: 261-282.

Islam, H. 2001. Sundarbans Forest Facing Serious Threats, *Wnwire*. United Nations Foundation. April 26, 2001.

Islam, K.L and M.M. Hossain (1986). Effects of ship scraping activities on the soil and sea environment in the coastal area of Chittagong, Bangladesh. *Marine Pollution Bulletin*, 17 (10): 462-463.

Islam, M.S. (2003). Perspectives of the coastal and marine fisheries of the Bay of Bengal, Bangladesh, *Ocean and Coastal Management*, Volume 46, Issue 8: 763-796.

K.U. (Khulna University) (1995). Integrated resources development of the Sundarbans reserved forest, Bangladesh. FAO,UNDP. BGD/84/056. 45 pp.

Karim, A. (1994). Environmental impacts. In. Z. Hussain and G. Acharya (eds), Mangroves of the Sundarbans. Volume II: Bangladesh. IUCN, Bangkok, Thailand. Pp. 203-218.

Katebi, M.N.A. and Habib, G. (1989). Sundarbans and forestry. *In* National workshop on coastal area resource development and management., Dhaka, pp. 79-100.

Khan, M.I, and Islam, M.R. (2003). Wastes Management in Offshore Oil and Gas: A Major Challenge in Integrated Coastal Zone Management. In: L. G. Luna (ed), CARICOSTA 2003 - 1st International Conference on Integrated Coastal Zone Management ICZM, University of Oriente, Santiago du Cuba, May 5-7, 2003.

Khan, M.I. and Islam, M. R. (2006a). Technological Analysis and Quantitative Assessment of Oil and Gas Development on the Scotian Shelf, Canada. *Int. J. Risk Assessment and Management*: *in press.*

Khan, M.I. and Islam, M.R. (2006b). Developing Sustainable Technologies for Offshore Seismic Operations, *J. Petroleum Science and Technology*: *in press.*

Khan, M.I. and Islam, M.R. (2006c). True Sustainability in Technological Development and Natural Resources Management. Nova Science Publishers, New York: 355pp.

Khan, M.I. and Islam, M.R., 2007. Handbook of Sustainable Oil and Gas Operations Management, Gulf Publish Company, USA. [accepted for publication in April, 2007].

Khan, M.I., Lakhal, Y.S., Satish, M., and Islam, M.R. (2006). Towards Achieving Sustainability: Application of Green Supply Chain Model in Offshore Oil and Gas Operations. *Int. J. Risk Assessment and Management*: *in press.*

Khatun, B.M.R. and M.K. Alam (1987). Taxonomic studies on the genus Avicinia Lin. from Bangladesh. *Bangladesh J. Bot.*, 16: 39-44.

Kumaraswamy, P.R. and Datta, S. (2006). "Bangladeshi gas misses India's energy drive?", *Energy Policy*: in press.

Lamparelli, C.C., F.O. Rodirigues and D.O. Moura (1997). Long term assessment of oil spill in a mangrove forest in Sao Paulo, Brazil. In: B. Kjerfve, L.D.de. Lacerda and E.H.S., Dipo (eds), Mangrove Ecosystem Studies in Latin America and Africa. UNESCO, Paris. Pp. 191-203.

Mahmood, N., S.Q. Chowdhury and S.Q. Sakait (1994). Indiscriminate expansion of coastal aquaculture in Bangladesh, genesis of conflicts: some suggestions. In: P.G. Wells and P.J. Ricketts (eds). Proceedings, Coastal Zone Canada '94. Cooperation in the Coastal Zone conference proceedings. *Coastal Zone Canada Association.*Vol.4, pp.1697-1706

Mukherjee, A.K. (1975). The Sundarbans of India and its biota. *J. Bombay Nat. Hist. Soc.* 72: 1-20.

Penn, J.W. (1983). An assessment of potential yields from the offshore demersal shrimp and fish stock in Bangladesh waters (including comments on the trawl fishery 1981/820. Rome, FAO, FI:DP/BGD/81/034, Field document 29p.

Saenger, P. and Siddiqi, N.A. (1993). Land from the sea: The mangrove afforestation program of Bangladesh, *Ocean and Coastal Management*, Volume 20, Issue 1: 23-39.

Saha, M., Sarkar, S.K., and Bhattacharya, B. (2006). Interspecific variation in heavy metal body concentrations in biota of Sunderban mangrove wetland, northeast India, *Environment International*, Vol. 32, No. 2: 203-207.

Sarker, M.S.U. and Sarker, N.J. (1988). Wildlife of Bangladesh. A systemitic list with status, distribution and habitat. Department of Zoology, University of Dhaka, 59 pp.

Silva, C.A.R. and A.A. Mozeto (1977). Relase and retention of phosphorus in mangrove sediments: Septiba Bay, Brazil. In: B. Kjerfve, L.D. de Laverda and E.H.S., Dipo (eds), Mangrove Ecosystem Studies in Latin America and Africa. UNESCO, Paris. pp. 179-190.

Stejskal, I.V. (2000). Obtaining Approvals for Oil and Gas Projects in Shallow Water Marine Areas in Western Australia using an Environmental Risk Assessment Framework, *Spill Science and Technology Bulletin*, Vol. 6, Issue 1: 69-76.

Teal, J.M. and R.W. Howarth (1984). Oil spill studies: a review of ecological effects. *Environmental Management*, 8 (1): 27-44.

Teresa, K.L., Wong, W.Y, Wong, Y.S., and Tam, N.F.Y. (2002). Fate of polycyclic aromatic hydrocarbon (PAH) contamination in a mangrove swamp in Hong Kong following an oil spill, *Marine Pollution Bulletin*, Volume 45, Issues 1-12: 339-347.

UN (2002). Born to Wild. The UN work for the environment. The United Nations, The UN Works Programme, Department of Public Information, Room S-955, New York, NY.

In: Sustainable Petroleum Operations
Editor: Rafiq Islam

ISBN 978-1-61324-157-8

Chapter 8

AIR POLLUTION AND CARBON ASSESSMENT FRAMEWORK FOR THE OIL AND GAS OPERATIONS

***M. Ibrahim Khan*[1]* *and S. Y. Lakhal*[2]**

[1]Institute for Sustainable Energy, Environment and Economy, University of Calgary
ES-602 University Drive NW, Calgary, AB, Canada, T2N 1N4
[2]Faculty of Business Administration
University of Moncton, Moncton, NB, Canada

ABSTRACT

Oil and gas operations causes numerous environmental problems. Air pollution is one of the major concerning issue in the whole operations. This paper examines the air pollution causes by the different oil and gas activities. Major air emissions are identified and these are: H_2S, natural gas, carbon monoxide, particulate maters, NO_x, volatile organic carbons, BTEX (Benzene, Toluene, Ethylbenzene, and Xylenes), Polycyclic aromatic hydrocarbons (PAHs), and CO_2. Among all of the pollutants, the CO_2 emission is the highest in quantity. As a result, special attention is given to assess the carbon. A framework is provided to conduct environmental assessment of Carbon as a valued ecosystem compound. A case study was carried out in the offshore oil and gas development on the

* Corresponding Author: Email: Ibrahim.Khan@dal.ca.

Scotian Shelf of Nova Scotia, Canada, to justify this proposed framework.

Keywords: offshore oil and gas; air pollution; carbon assessment; valued ecosystem component; environmental impact assessment; Scotian Shelf.

1. Introduction

Oil and gas operations are involved numerous activities in different phases. Different phases of oil and operations exploration, drilling and installation, production, transportation, refining and processing, and decommission. Every phases and activities of oil and gas operations are associated with numerical wastes and toxic releases into air, water, and land (Khan and Islam, 2007). This main objective of this paper is to examine air pollution from the oil and gas operations. Until now, the petroleum oil is the major energy sources and due to its high market demand the production and explorations has been increased in the all around the world. As a result, the wastes generation and pollution releases especially the air pollution has been reported as major concern in upstream and downstream oil and gas operations. According to Environmental Integrity Project (EIP, 2007) the air pollution from oil and gas operations are responsible for the increase of cancer in USA. According to EIP, only nine oil refineries in the United States account for only 15 percent of the nation's refining capacity, but a third of total cancer-causing emissions reported by the oil industry. The carcinogens emitted by U.S. refineries include benzene, ethylbenzene, butadiene, polycyclic aromatic hydrocarbons, naphthalene, formaldehyde, and metals such as nickel and lead (EIP, 2007).

2. Air Contaminants

Different types of air emissions are taken place from the exploration, production, and processing of oil and gas. Figure 1 pictorially presents the different activities and associated wastes generation in the different phases of oil and gas operations. Major air emissions are: sour gas, flaring and venting, benzene emissions, and other air emissions. Table 1 shows the major air emissions from different activities associated with oil and gas production.

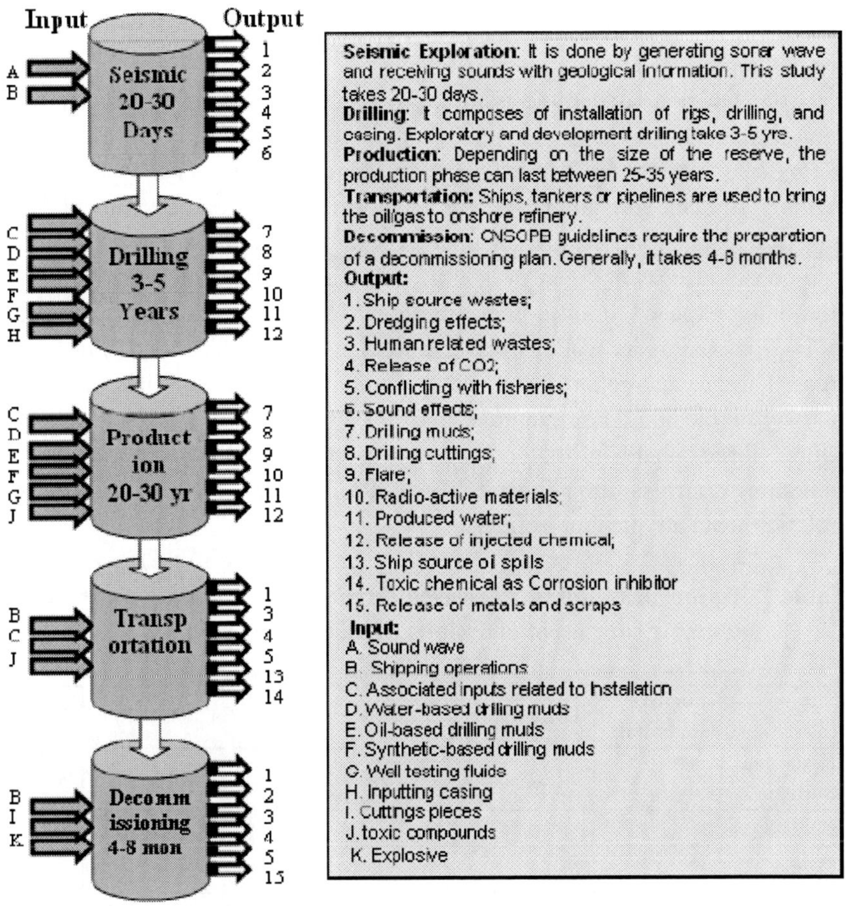

Figure 1. Pollutants and wastes released in the different oil and gas operation phases.

2.1. Sour Gas

Sour gas is hydrogen sulphide (H_2S) contains natural gas. The sour gas smells as rotten egg due to the presence of H_2S. The odour of H_2S is strong and can be detectable in the air at very low concentrations, such as 10 parts per billion (0.000001 per cent). Sour gas contains over one per cent H_2S. If the gas contains more than one per cent H_2S, more corrosion-resistant materials and equipment must be used to produce gas safely. Exposure to the H_2S can be toxic and lethal to the human, wildlife and farmed animals if inhaled at high concentrations. If the concentration is more than 10 parts per million, then workers require special training and equipment on worksites where the natural gas being produced or processed (Bott, 2006).

Sour gas is a concerning issue in West Canada, because one third of the natural gas produced in this region is considered sour, (CAPP, 2006). H_2S may be toxic to humans and animals in low concentrations. H_2S's potential for harming human and animal health. According to U.S. government health statistics, concentrations greater than 500 parts per million can cause you to lose consciousness and can even kill you. Children and the elderly may die after being exposed to lower concentrations (CBC, 2005). Most of those killed by H_2S poisoning are workers exposed to high levels in confined, unventilated areas. There is also report that some workers have also been killed during sour gas well explosions. For example, in 2005, an explosion killed one man and injured three others in southern Alberta as they tried to bring an oil well containing sour gas into service (CBC, 2005). Figure 2 shows the scenario of emission from a typical oil refinery.

Table 1. Major air pollutants released during oil and gas development and the major sources of emissions (Source: Earth Works, 2003)

	Fugitive Emissions	Dehydration	Vessels	Flaring	Engines	Pits	Venting
Hydrogen Sulfide	x			x			x
Methane	x	x				x	x
Ozone	o	o	o		o		
Carbon Monoxide			x	x	x		
Nitrogen Oxides			x	x	x		
VOCs	x	x		x	x	x	x
BTEX	x	x		x	x	x	x
Particulate Matter		x	x		x		
BTEX	x	x		x	x	x	x
PAHs				x			

'x' means that a pollutant is emitted as a direct result of the particular activity, and 'o' means that the pollutant is generated in a secondary reaction associated with the particular oil and gas development activity.

Figure 2. Pollutants and wastes released in the different oil and gas operation phases.

2.2. Flaring and Venting

In the reservoir, different natural gases are found with oil. When oil comes out from reservoir these gases are expands and releases and these gases are known as solution gases. These solutions gases are other gases which are considered as unviable due to economic and technical reasons are released or burned out. Flaring is a technique has been used in the petroleum industry to burn out unwanted gas. Figure 3 shows the gas flaring in oil refinery. There is another technique which is known as venting, in this process the undesired gases are released in the air.

The flaring and venting are well established techniques in production and processing of crude oil, but these are most concerning issues. These techniques are directly related to air pollutions. In addition, these techniques are also responsive for the destructions of enormous amount of valuable natural resources of a nation. The low quality gas that is flared contains many impurities and during the flaring process, toxic particles are released into the atmosphere. Acid rain, caused by sulfur oxides in the atmosphere, is one of the main environmental hazards resulting from this process.

Figure 3. Flaring from a typical oil refinery, Atlantic Canada.

It is reported that due to unsustainable exploration practices and lack of gas utilization infrastructure the flaring rates is very high. For example, in Nigeria, the country flares 75 per cent of the gas it produces (Evoh, 2002).

According to UKOOA (2002), the flaring of natural gas accounts for approximately a quarter of the petroleum industries emissions. Flaring also contributes 172,208 thousands metric tons of CO_2 releases in the atmosphere in 1998 (Earth Trend, 2003). To overcome this problem Bjorndalen et al. (2003) proposed a no-flare design and they elaborated various options in achieving a no-flare process. Khan and Islam (2007) provided sustainable technology development to avoid flaring and venting in the petroleum operations.

2.3. Benzene Emissions

In the oil and gas operations benzene may released into atmosphere from point sources (CAAP, 2006). Benzene is a light aromatic hydrocarbon and highly water-soluble compounds in BTEX and PAH group and considered as higher risk quotient value (Berry and Wells, 2004; Khan and Islam, 2005). It occurs naturally as a part of raw oil and natural gas and considered to be a non-threshold carcinogen and is both an occupational health and public health issue (CAAP, 2006). In addition to benzene, toluene, ethylbenzene, and xylene are also reported to emit (Earth Trend, 2003), which are know as BTEX. The major sources of BTEX emission are: incomplete combustion of natural gas in flares, venting, fugitive emissions from stock tank vents and sumps leaks in processing equipment, glycol dehydrators additives, engines, produced water storage tanks, and during the dehydration of natural gas.

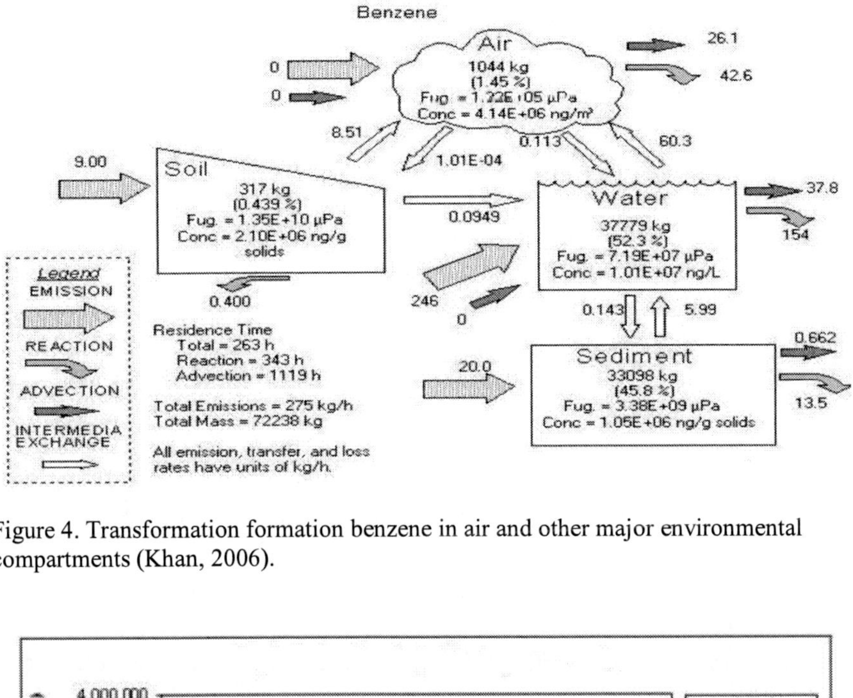

Figure 4. Transformation formation benzene in air and other major environmental compartments (Khan, 2006).

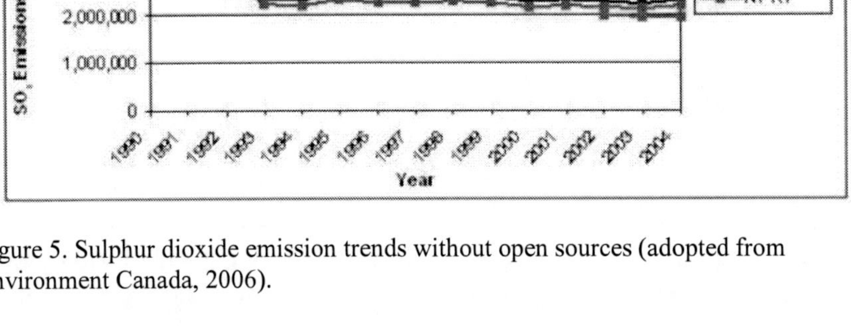

Figure 5. Sulphur dioxide emission trends without open sources (adopted from Environment Canada, 2006).

Figure 4 shows the transformations of benzene into air and other environmental components. Benzene is a known carcinogen, and has also been shown to cause blood disorders and to impact the central nervous system the reproductive system. Toluene may affect the reproductive and central nervous systems. Ethylbenzene and xylene may have respiratory and neurological effects.

2.4. Sulphur Dioxide

Oil and gas operations can emit sulphur dioxide (SO_2). They are from different sour gas production, flaring at oil facilities, well-testing flares, gas plants, and oil sands facilities. This emitted SO_2 combines with water and causes acid rain or acidifications. SO_2 reacts with other chemicals to form particulate pollution, which can damage lungs and cause respiratory illness, heart conditions, and premature death.

Figure 5 shows the SO_2 emission trends in Canada. After 1993 the SO_2 emissions was dropped significantly, but the emission from oil and gas operations has increased significantly (Environment Canada, 2006).

2.5. Carbon Monoxide

Carbon monoxide (CO) is emitted during flaring and from the operation of various machinery at oil and gas development sites. Table 1 shows the emission of CO from the three different activities of oil and gas operations, which are vessels, flaring and engines. It is a colorless, odourless, flammable gas. Due to incomplete burning of carbon-based fuels such as, coal, and even wood it is produced. CO is poisonous if inhaled. It inhibits the blood's ability to carry oxygen, and can cause dizziness, unconsciousness, and even death (Earth Works, 2003).

2.6. Natural Gas

During the oil and gas operations natural gas can be emitted. It might be released from different activities, such as accidentally, leaking from equipment, and in venting operations. The primary component of natural gas is methane, but there are other minor components, such as ethane, propane, butane, and pentanes. Hazardous air pollutants such as benzene, toluene, ethylbenzene, xylenes and hexanes, H_2S, and CO_2 are also found in the natural gas. In addition to hazardous pollutants, there are other compounds in natural gas typically include water vapor, helium, and nitrogen. In table 1, the releases of primary components of natural gas methane is shown.

2.7. Nitrogen Oxides

Nitrogen oxides (NOx) are formed during the combustion of fossil fuels, which causes a chemical reaction between nitrogen (which occurs naturally in the atmosphere) and oxygen. During oil and gas production, NOx are formed during flaring operations, and when fuel is burned to provide power to machinery such as compressor engines and other heavy equipment. NOx, in turn, may react with VOCs to form ground-level ozone.

The health impacts from NOx include respiratory problems, heart conditions, and lung damage. Figure 6 shows the emissions trends of NOx in Canada. It is relatively stable over the past 15 years, it is reported that increases from certain oil and gas operations and electric power generating facilities using fossil fuels (Environment Canada, 2006).

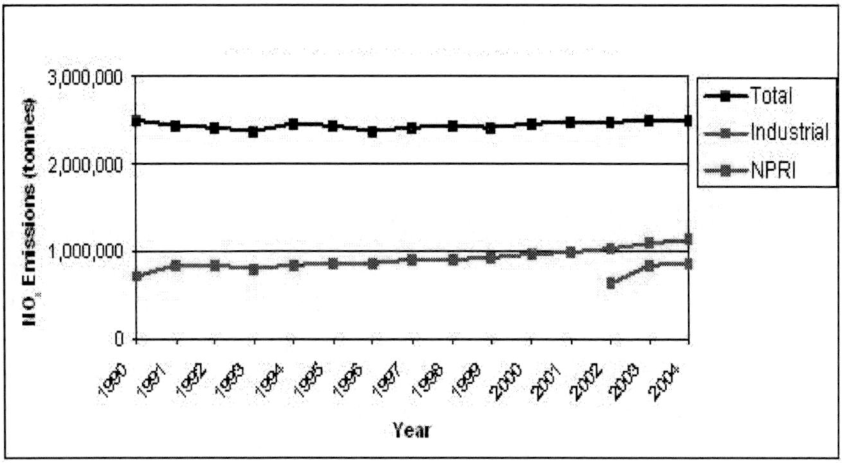

Figure 6. Nitrogen oxide emission trends without open sources (adopted from: Environment Canada, 2006).

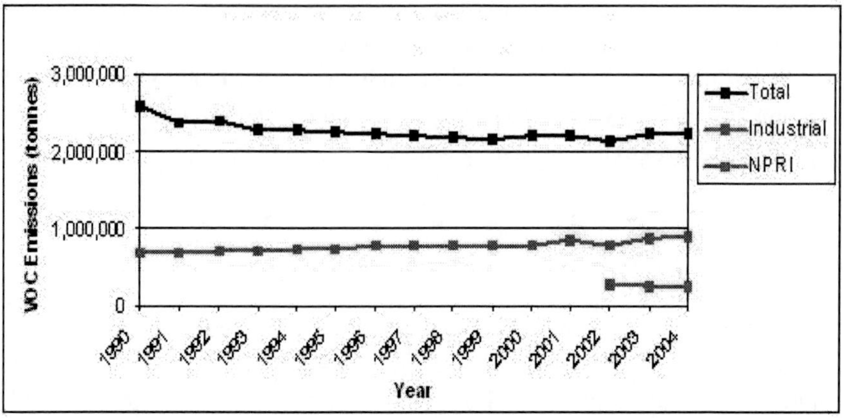

Figure 7. Volatile organic compounds emission trends without open sources (adopted from Environment Canada, 2006).

2.8. Ozone

In the oil and gas operations, ozone itself is not released, but some of the main compounds that combine to form ozone are released. They are: volatile organic compounds and nitrogen oxides. Ozone, when found at ground-level, is also referred to as "smog," which, when inhaled can cause or aggravate respiratory ailments such as asthma (Earth Work, 2003).

2.9. Particulate Matter

Oil and gas operations can release particulate matter (PM). The PM is small particles (i.e., dust) released in the air from different oil and gas activities, such as flaring and venting, diesel exhaust from engines and power machinery, dust or soil entering the air during pad construction. The released PM is gradually settled down to the ground. It causes numerous health problems when inhales. According to Earth Works, 2003) it is reported that particles may lead to adverse health effects such as respiratory or breathing ailments, cancer, or premature death depending on the composition of PM.

Table 2. Carbon dioxide emission from energy and industry in Canada (data source: Earth Trends, 2003)

Carbon Dioxide (CO_2) Emissions (in thousand metric tons of CO_2)	Canada	North America	World
Total Emissions, 1998	467,223	5,915,907	24,215,376
Percent change since 1990	9%	13%	8%
Emissions as a percent of global CO_2 production	1.9%	24.4%	-
solid fuels	103,508	2,085,772	8,654,368
liquid fuels	192,818	2,467,484	10,160,272
gaseous fuels	160,842	1,298,217	4,470,080
gas flaring	4,044	16,599	172,208
cement manufacturing	6,012	47,835	758,448

2.10. Volatile Organic Compounds

Oil and gas operations activities can release the volatile organic compounds (VOCs). The VOCs are carbon-containing substances that readily evaporate into the air and combine with nitrogen oxides to form ground-level ozone. Common VOCs from oil and gas operations are benzene and toluene. It is reported that VOCs cause respiratory ailments such as asthma, and decreased lung function (Earth Works, 2003). Figure 7 shows the emissions of VOCs in Canada. It decreased slightly between 1990 and 1999, with the most notable decrease occurring from 1990 to 1993. However, according to Environment Canada (2006) in the hydrocarbon sectors the VOCs have been increased. The oil and gas operations are one of the main sources of VOC emissions reported to the NPRI since 2002 (Environment Canada, 2006).

3. Greenhouse Gas Emissions

Atmospheric CO_2, along with two other gases, nitrous oxide and methane, which are all commonly referred to as 'greenhouse gases', act to increase the heat retention capacity of the earth's atmosphere (Bouwman, 1989). The increase in greenhouse gas production by human activities has been argued to be partially responsible for average global temperature increase, which in turn, may bring on shifts in many natural equilibria, with great significance to biological systems (Houghton, 2001).

The significance of atmospheric carbon levels to global environmental conditions, and its close link with oil and gas development through the combustion of petroleum products makes atmospheric carbon an important valued ecosystem component (VEC). There is no evidence to suggest that a unique relationship between oil and gas development and atmospheric carbon exists in particular area. This work, therefore, discusses the subject in general and global terms. This study investigates impacts petroleum operations on carbon VEC and case study has been taken in regards to offshore oil and gas operations in Nova Scotia, Canada.

4. Assessment of Carbon as a Valued Ecosystem Component

The risks of climate change are pressing the oil and gas industry to minimize its greenhouse gas emissions especially CO_2, because it CO_2 from the burning of fossil fuels is the largest single source of greenhouse gas emissions from human activities. Table 2 shows the total amount of CO_2 from Canada, North America and whole world. The supply and use of fossil fuels accounts for about 80 percent of mankind's CO_2 emissions, one fifth of the CH_4, and a significant quantity of N_2O (UNFCC, 2006).

Most emissions associated with energy use result when fossil fuels are burned. In this study, exploration, development, processing, transporting, and distributing oil and gas were investigated for greenhouse gases releases. These releases can be deliberate, as when natural gas is flared or vented from oil wells, emitting mostly carbon dioxide and methane, respectively. They can also result from accidents, poor maintenance, and small leaks in well heads, pipe fittings, and pipelines (UNFCC, 2006).

Now the challenge is to assess the environmental impacts on carbon VEC in regards to oil and gas operations. For analyzing the status and effects levels on carbon different issues such as, its distributions, policy commitment, applicable monitoring protocol, typical pre-mitigation are discussed. Figure 8 shows the detailed issues we considered in this study.

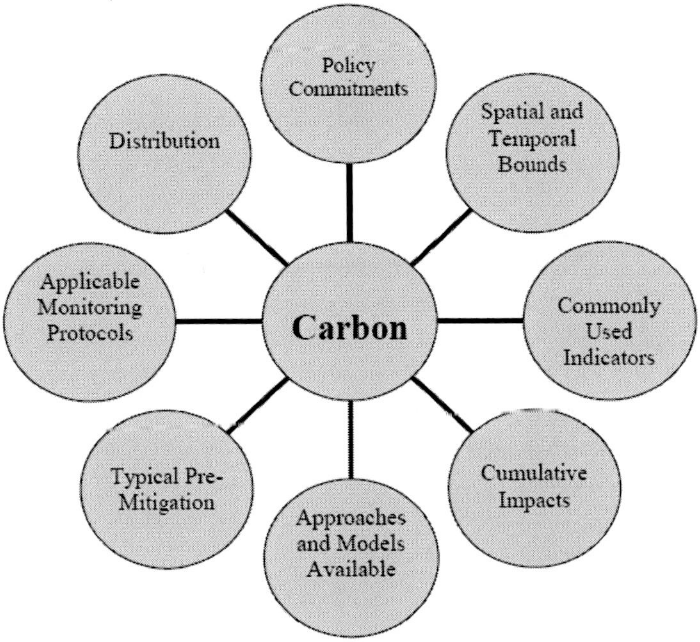

Figure 8. Multiple factors of carbon valued ecosystem components.

4.1. Policy Commitments Related to Carbon

The Government of Canada, by signing the Kyoto Protocol has committed itself to limit and reduce emissions of six greenhouse gases (including carbon dioxide) during the period 2008-2012. Canada has agreed to limit its average annual emissions during the period 2008-2012 to 94% of its baseline emissions, a reduction of 6% from the base line and 20-30% from the projected emissions in 2010 (Haites, 2000). In October 2000, the province of Nova Scotia along with the other provinces and the federal government accepted a national implementation strategy on climate change (Government of Nova Scotia, 2001).

According to the Canadian Association of Petroleum Producers the industry is only responsible for 20% of the total greenhouse gas emissions in Canada during the production and distribution of oil and gas with the other 80% attributed to end-use consumption (CAPP, 2002). The government of Nova Scotia estimated that, in 1998, 20100 kilotonnes (kt) of CO_2 was emitted by the province as well as projecting an increase to 21100 kt by the year 2010.

This model suggests that if Nova Scotia plans to achieve Kyoto's target the province will have to reduce emissions by 16% (Government of Nova Scotia, 2001).

The CNSOPB is responsible for regulating CO_2 emissions, which result from oil and gas activities in offshore Nova Scotia. The new draft of CNSOPB regulations, Section 2(2.2) states that companies pursuing any type of activity in the offshore should "provide an estimate of annual quantities of greenhouse gas (GHG) that will be emitted from its offshore installation(s) and a description of plans to control and reduce these emissions" (CNSOPB, 2002).

It should also be noted that the Canadian Association of Petroleum Producers (CAPP) has made a voluntarily commitment towards efforts to reduce GHG emissions from operational facilities. Since 1995 most oil and gas companies in Canada have voluntarily participated in Canada's Voluntary Challenge and Registry (VCR) program, where companies are encouraged to limit their net GHG emissions and report to CAPP annually on their progress towards achieving such a goal (CAPP, 2002).

4.2. Spatial and Temporal Bounds for Impact Assessment

The selection of time frame and spatial scope are critical in accurate EIA (Peterson, et al. 2001) particularly in the 'widening' and toughening of environmental regulation (Salter and Ford, 2001). It is not the impact at the local level on atmospheric carbon concentration, but the cumulative effect of many sources at the global scale, that presents environmental problems by increasing atmospheric carbon levels. This potentially acts to increase adverse effects on humans through global warming (Houghton, 2001).

Because atmospheric carbon rapidly dissipates into the atmosphere, the impact of one petroleum development project will not be easily observed. Further, seasonal and annual fluctuations of atmospheric carbon concentrations require a broader analysis to find overall trends on a global scale (Esser, 1989). It is therefore recommended that the approach taken is to assess global capacity to absorb atmospheric carbon and budget allowable contributions accordingly. Offshore petroleum activities, which release atmospheric carbon are the combustion of fuels during exploration, extraction and decommissioning, and the burning off of undesired petroleum products (flaring) extracted along with desired product.

Measuring impacts of fuel combustion will provide a basis to estimate the amount of atmospheric carbon released by a given activity. A considerable contribution to atmospheric carbon levels arises from the combustion of the extracted product (Houghton, 2001), but conventional EIA typically does not consider the impacts of the use of a product resulting from an activity. However, in the interest of considering the full implications of petroleum development on atmospheric carbon levels, there are grounds to integrate the impacts of product use into the overall assessment. The relative contributions to increased atmospheric carbon levels of extraction activity compared to consumption of product will increase as more easily accessible fuel reserves are exhausted.

One aspect of the current EIA framework, which may provide answers to the challenge of integrating product-use impacts is the preliminary stage of assessment, which involves determining the need of, and alternatives to a given project. Cost benefit analysis provides a means to weigh the overall benefit of a given activity against other alternatives, such as alternative energy sources, which contribute less to global warming. A strict assessment of cumulative impacts of any atmospheric carbon-producing activity would likely preclude that activity on the basis of current atmospheric carbon trends. Thus considering cumulative effects would not useful because of the potential preclusion of a significant segment of industrial activity in the current social and economic context.

Life cycle analysis (LCA) was developed to assess the environmental impacts of a product, process or activity over its entire life cycle (Salter and Ford, 2001). Its application can be used to broaden the EIA scope to include product consumption. No examples of this approach were found, although Itsubo and Yamamoto (1999) provide an example of inclusion of broad impacts such as global warming among impact categories in LCA of different technology options for a given activity.

Generally, LCA does not merge fine detail well with larger impacts, although it has some potential to do so with improvement in the use of fate models in marine environments. Such holistic environmental assessment techniques help to overcome the failures of project-level EIA in assessing total environmental risk (Salter and Ford, 2001), however further development in integrating product consumption into EIA should be pursued.

4.3. Commonly Used Indicators for Impact Assessment

Atmospheric carbon is released from offshore oil and gas facilities through several activities. These include the production of power on rigs by the burning of fossil fuels, flaring of oil and gases from drilling activities, and water and gas injection that plays a role in extracting oil. The goal of discussing the commonly used indicators for atmospheric carbon as a VEC is to be aware of how we can assess the status of this VEC. What needs to be established is how offshore oil and gas projects throughout their lifecycle produce carbon that goes into the atmosphere. This is done through combustion of fossil fuels throughout the life of the project to produce power on the rig and support vessels, and with drilling that requires flaring (Patin, 1999). In order to assess the levels of carbon being emitted the total amount of fossil fuels that have been combusted in any manner must be known. The indicators for this aspect of the process include the careful tracking of fossil fuel use and the amount of hydrocarbons being flared off during drilling stages. Once the volume of hydrocarbons combusted by the project over a specific time period (e.g. yearly) is known then information regarding the amount of carbon released into the atmosphere through combustion can be established. Through these methods the amount (tons of CO_2 emitted) of carbon emitted by a project can be established and then compiled to establish total carbon emissions by sector (e.g. offshore oil and gas exploration and production off Nova Scotia).

The measurement of consumption of hydrocarbons by combustion for power generation throughout the project is done by accounting for the amount used from storage tanks. This data is then translated depending on the type of hydrocarbons burned into carbon emissions levels. The other source of atmospheric carbon from these projects is flaring. The ability to account for the amount of hydrocarbons being combusted in flaring is made possible by hydrocarbon flow meters on flaring systems (Hussain et al. 2001). Measured flow rates can then be translated into hydrocarbon burn-off amounts, which can then be translated into carbon emissions levels.

Other indicators of offshore oil and gas production's contribution to atmospheric carbon levels can be seen in localized atmospheric fallout of hydrocarbon residues (Patin, 1999). Less direct indicators include increased localized atmospheric carbon levels above background levels and the contribution of the industry to global atmospheric carbon emissions.

4.4. Approaches and Models Available for Impact Prediction

In addition to other air emissions, carbon, in the form of carbon monoxide (CO), carbon dioxide (CO_2), and some methane (CH_4) is emitted from engines on the drilling platforms, vessel and aircraft engines, flaring, and other fugitive emissions (e.g. spills). Project malfunctions or accidents can lead to relatively minor and temporary discharges of SO_2, H_2S, and CO_2.

There are a number of biophysical and chemical processes, which influence oil (and indirectly, carbon) properties, including early and long term weathering processes, and processes that affect the location of oil in the water column. Early weathering processes that change oil properties include: evaporation, dispersion, dissolution, and emulsification. Long-term weathering processes include: biodegradation, oxidation, and sedimentation. Weathering processes that change the location of oil include: advection, spreading, submersion, and stranding (Khan, 2006).

Many models exist that can be used to measure impacts of the oil and gas industry's carbon emissions on carbon levels both on the atmosphere and the surrounding ocean. In addition to global atmospheric CO_2 models (e.g. The Integrated Model to Assess the Greenhouse Effect <IMAGE>), there are also models that describe impacts at local levels, for example, from the Sable Island Offshore Energy Project. The following models describe a variety of impacts on carbon levels, including atmospheric and water dispersion, hydrocarbon dispersion, the probability of different sized oil spills, and drill cuttings dispersion. In this section a selection of models will be described including their overall functions, and the necessary inputs required to run the model/simulations.

Atmospheric Dispersion

Hadley Ocean Carbon Cycle (HadOCC) Model

On entering the ocean, carbon dioxide undergoes rapid chemical reactions with the water and only a small fraction remains as carbon dioxide. The carbon dioxide and the associated chemical forms are collectively known as dissolved inorganic carbon or DIC (Met Office, 2001). This model describes the movement of carbon, in its various forms throughout the global ocean carbon cycle, from reactions at surface waters, transport within the global ocean current, to being taken up by various organisms within the ocean.

Integrated Model to Assess the Greenhouse Effect (Global)

IMAGE is a parameterized simulation model, developed for the calculation of historical and future emissions of greenhouse gases on global temperature and sea level rise and ecological and socio-economic interests in specific regions. It is a deterministic computer simulation model, which consists of interlinked modules, (the output of one module serves as the input for the next), which each describes a specific element of climate change (Rotmans, 1990). Two of the modules are the Emission Module and the Ocean Module. The Emission Module includes the trace gases CO_2, CH_4, CO, N_2O, CFC-11, and CFC-12 and their emission from different sources grouped as: nature, energy, agriculture, and industry. The Ocean Module shows the ocean has being divided into 12 layers. Three primary processes are defined, determining the carbon concentration in the ocean: carbon transport by mass flow of water, by turbulent mixing, and by gravitation. This is a one-dimensional "box-diffusion" model, which does not yet include polar outcrop, nutrient cycling, upwelling and differences between the Atlantic, Pacific, and Indian oceans. Although one-dimensional box-diffusion models are useful for general analysis of the role of the oceans in the global carbon cycle, they do not give a realistic representation of the dynamics of the ocean (Rotmans, 1990).

Industrial Source Complex Dispersion Model (Local)

ISC (Industrial Source Complex Model) has been specifically developed to simulate air pollution due to an industrial plant, taking accurately into account the effect of high stacks on the behaviour of the pollutant plume. This could inadvertently be used to model CO_2 emissions also. Numerous options include the ability to compute the dry deposition of the pollutant downwind of the stacks, to model the plume height accounting for the hydrodynamic effects, and to simulate the impact of linear, area and volumetric sources (GAIA, 2002). It works with non-reactive pollutants, including particulate matter. The ISC-PRIME model is a steady-state Gaussian plume model, which can be used to assess pollutant concentrations from a wide variety of sources associated with an industrial source complex (GAIA, 2002). This model can account for the following: settling and dry deposition of particles; building downwash; area, line, and volume sources; plume rise as a function of downwind distance, building dimensions and stack placement with respect to a building; separation of point sources; and limited terrain adjustment. ISC-PRIME was used by consultants in PanCanadian's Environmental Impact Statement for their proposed development of the Deep Panuke oil fields.

Water Dispersion Models (Developed by DNV Environmental Consulting)

Oiltraj

This model is an offshore oil spill model integrated in an interactive GIS (Geographical Information System) application, which means that a model of the oil spill can be displayed as part of a map. It calculates the displacement and mass balance of oil particles released from a fixed position. Wind and sea surface currents drive the particles. The mass balance calculation includes effects such as evaporation, natural dispersion (oil mixed down into the water masses) and emulsification (water uptake) (Det Norske Veritas [DNV] 2000).

Shoredrift

SHOREDRIFT is a numerical oil drift model developed for the simulation of coastal zone spills. This model is also integrated in an interactive GIS application. The model can run in statistical mode as well as single scenario mode. Necessary input data are wind speed and direction, current observations, type of shoreline, oil characteristics and release scenario (DNV, 2000). The statistical mode of the model computes probabilities of pollution, mass budgets (for evaporation, down-mixing {natural dispersion}, water uptake and remaining oil mass) and drift times for the entire computation area (DNV, 2000).

Oilslick/DIFF

OILSLICK and DIFF are tools for computing hydrocarbon (HC) concentrations in water masses and for weathering studies. The two models may be run together or separately. OILSLICK computes the behaviour and characteristics of an oil slick i.e. thickness, width and budgets for evaporated, dispersed and dissolved fractions (as a function of the distance from the discharge point) of HC, generated from a surface spill (DNV, 2000). DIFF computes the HC concentrations in water masses, following a surface or a subsurface release (blowout).

BLOW

BLOW is applied for near field simulations of subsea releases of gas and oil (blowout). A general feature of an underwater blowout is that oil and gas, under large amounts of pressure, are released, followed by an intense mixing of oil, gas and water masses. This model calculates plume characteristics such as the mixed amount of seawater, plume speed and plume diameter for various depths (DNV, 2000). The input data are typically surface currents, where the plume is expected to reach the surface, vertical temperature and salinity profiles, and release characteristics.

DRILL

The DRILL model is used for calculating dispersion and seabed sedimentation of drill cuttings and muds from offshore installations. The user specifies the amount of drill cuttings discharges and the density and particle size distribution of the cuttings (DNV, 2000). It is assumed that the discharge takes place over a long time period (> 1 year) for the statistics to be valid. Information on ocean currents is input as a frequency or probability distribution of current speed and direction at one or several depths.

4.5. Typical Pre-Mitigation Impacts

The majority of atmospheric emissions of atmospheric carbon occur during the exploratory drilling and production stages of oil and gas. Figure 8 show the amount of CO_2 emission for oil and sector. According to Patin (1999) the main sources of these emissions include:

- Constant or periodic burning of associated gas and excessive amounts of hydrocarbons during well testing and development
- Continuous flaring to eliminate gas from the storage tanks and pressure-control systems
- Combustion of gaseous and liquid fuel in the energetic units (diesel-powered generators and pumps, gas turbines, internal combustion engines) on the platforms, ships, and onshore facilities
- Evaporation or venting of hydrocarbons during different stages of production, treatment, transportation, and storage

Kwon and Schnoor (1994) suggest that if CO_2 emissions from fossil fuel combustion continue at the present increasing rate of close to 1.5% per year, a CO_2 doubling (to 560 ppm) will appear by the year 2060. Kessel (2000) estimates that continued use of fossil energy will lead to an increase in the average global temperature by 1.0-3.5 ^{o}C in the coming 50-100 years. Though these and other forecasts of future CO_2 emissions from fossil energy use as well as the magnitude of their influence on global warming are highly disputed, the impact of CO_2 emissions on global warming itself is widely accepted (Kessel, 2000).

Possible impacts of climate change and variability on marine life might include changes in living conditions via changes in water temperature, wind and waves and other physical climate related forcing of the marine environment (Haugan, 1998). There is also a chemical impact on the marine environment caused by atmospheric CO_2 emissions penetrating into the oceans (Haugan, 1998). Increased levels of atmospheric CO_2 will lead to increased invasion across the air-sea-interface, and could lead to climate change but could also be harmful to marine food webs and ecosystem functions (Haugan, 1998).

Since the industrial revolution, the oceans have served as a sink for 30% of all anthropogenic CO_2 (Raven and Falkowski, 1999). This is mainly due to CO_2 solution in the surface ocean; however, a poorly quantified fraction is a result of increased production due to anthropogenic inputs of combined nitrates, phosphates and iron (Raven and Falkowski, 1999). 'Circulation as usual' models predict that future anthropogenic inputs of CO_2 into the atmosphere will, in part, continue to be sequestered in the ocean (Raven and Falkowski, 1999). However, changes in oceanic thermohaline circulation as a result of global climate change would greatly alter the predictions of carbon sequestration that are possible on a 'circulation as usual' basis (Raven and Falkowski, 1999). Also, ocean-buffering capacity to take up the excess CO_2 will decrease with the increasing atmospheric CO_2 concentration (Kwon and Schnoor, 1994).

4.6. Typical Mitigation Measures and Residual Impacts

Although atmospheric emissions accompany most of the oil and gas operations, they have thus far not received adequate attention, probably due to the remoteness of offshore developments (Patin, 1999). However, the technical means to mitigate atmospheric pollution during offshore oil and gas production exist and are analogous to methods used on land and in other industries. Mitigation of increased levels of atmospheric carbon emissions can be accomplished in two main ways: carbon sequestration and reduction of emissions.

Artificially Enhanced Atmospheric Carbon Sequestration

Carbon sequestration is the accumulation and storage of carbon in locations other than the atmosphere. The storage sites, or sinks, range from oceans and forests to agricultural soils and cover both natural and human-made stocks (Sedjo, 1998). Because climate change will reduce the ocean's efficiency at sequestering atmospheric carbon, it has been proposed that human intervention could increase carbon sequestration in the oceans (Raven and Falkowski, 1999). This intervention would include large-scale ocean fertilization to increase carbon sequestration by phytoplankton and/or direct CO_2 burial in the deep ocean.

Ocean Fertilization

The stimulation of primary production can be correlated with the increased inputs of nutrients that limit productivity (Haugan, 1998). Iron directly enhances primary production in high nutrient (nitrate and phosphate) regions of the oceans (Raven and Falkowski, 1999). This increase in primary productivity will lead to an increase in sequestration of carbon by phytoplankton, and hence by the oceans in general (Haugan 1998). It should be noted that this type of fertilization could also lead to unpredicted environmental perturbations (Raven and Falkowski, 1999).

CO_2 Burial

New and improved technology will allow substantial volumes of CO_2 to be separated from gas turbine exhausts, CO_2 compression and pipelines and then deposited deep beneath the seabed in old wells or underground aquifers (Aunan, 2001; UNEP, 2002). In September 1998, Statoil, the largest Norwegian oil company initiated a pilot program to separate CO_2 from the waste products of power production and subsequently inject the CO_2 into the seabed of the North Sea (Energy Information Administration 2002). This mitigation measure may have the greatest potential emissions reductions (Aunan, 2001) but according to Statoil, it is still not cost effective (Energy Information Administration, 2002). It should also be noted that, according to Haugan (1998), the direct storage of CO_2 in the seabed might, on a local scale, impose a stronger disturbance to the marine environment than indirect storage via the atmosphere. This disturbance may be caused by CO_2 leaking out from the seabed into the water column above (Haugan, 1998).

Reduction of Emissions at Platforms

Flaring

The flare is a security measure, and traditional flaring entails that a platform has a continually burning flare that makes it possible to quickly burn off any excess gas from the processing (Patin, 1999). One way to reduce the amount of atmospheric carbon released in the offshore is to use cold flaring. Cold flaring involves using a flare that is lit automatically as the need arises, combined with changes in the production process so that the need for flaring is reduced in general (Aunan, 2001). However, despite the fact that most flaring can be eliminated, there will always be the need for some flaring for safety purposes (Patin, 1999).

Enhanced Oil Recovery

Using CO_2 for pressure support to improve oil recovery is also under consideration (UNEP, 2002). Norsk Hydro, Norway's second largest oil company, has investigated the effectiveness of capturing CO_2 from flue gas and subsequently injecting the CO_2 into oil wells (Energy Information Administration, 2002). This process, known as enhanced oil recovery, not only sequesters CO_2 (thereby reducing emissions of this greenhouse gas into the atmosphere), but also increases the quantity of oil that can be retrieved from wells (Energy Information Administration, 2002).

Emission Taxes

Norway is currently the only country that imposes emissions taxes on the release of atmospheric carbon in offshore operations (Energy Information Administration 2002). However, the use of such taxes could be used to promote the reduction of flaring, and to encourage efficiency of processing activities in Nova Scotia (National Climate Change Secretariat , 1998). However, some companies may chose to continue emitting high levels of atmospheric carbon and pay the taxes in which case this method will not help to lower emissions.

4.7. Applicable Monitoring Protocols

Carbon dioxide (CO_2) emissions are largely dependent on the carbon balance of a fuel and do not vary widely by combustion process, while carbon monoxide (CO) emissions are largely dependent on combustion and atmospheric conditions, and they are both influenced substantially by emission control technologies (OECD, 1991). Therefore, the desirable energy data to estimate the amount of these gases produced varies from simply primary fuel requirements for CO_2, to very detailed energy end-use and pollution control data for CO. An emission factor is generally defined as an average quantity of emission per unit of emitting activity (OECD, 1991).

Emission factors are averages derived from a number of data points of monitored emission levels from a single type of activity or technology being used in different places under different conditions. Emissions from stationary energy systems (e.g. engines on the platforms) are usually measured by mass (g) or pollutant present in the flue gas, given a particular oxygen content of the flue gas. If the activity is energy production or consumption, it can be measured in fuel weight (mass) or volume per energy input or energy output. For CO_2, the coefficient of emission per unit of energy combusted is determined by a mass balance calculation (OECD, 1991). It is assumed that the total carbon content of a fuel is passed into an equivalent mass of carbon in CO_2 emissions. It is important to realize that, because CO and CH_4 are small in mass by comparison to CO_2, they can be reasonably ignored without affecting the global or regional estimates of CO_2 (OECD, 1991). Figure 9 shows the different pollutants releases during oil and gas operations.

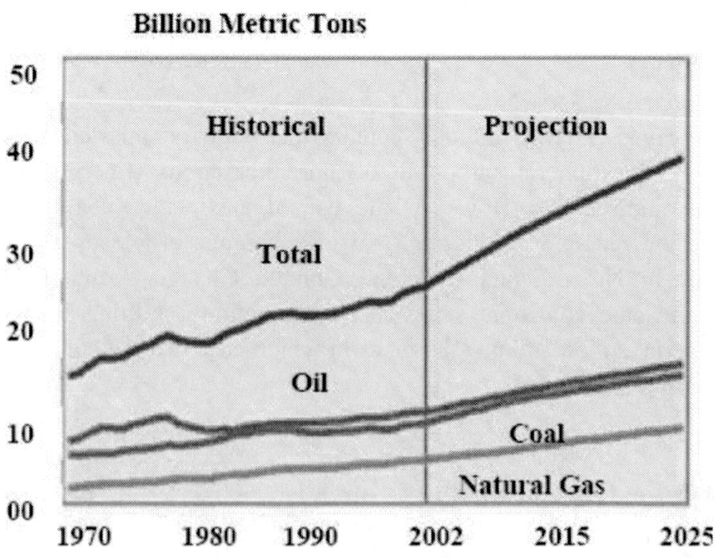

Figure 9. World CO_2 Emissions by oil, coal and natural gas, 1970-2025 (adopted from EIA, 2004).

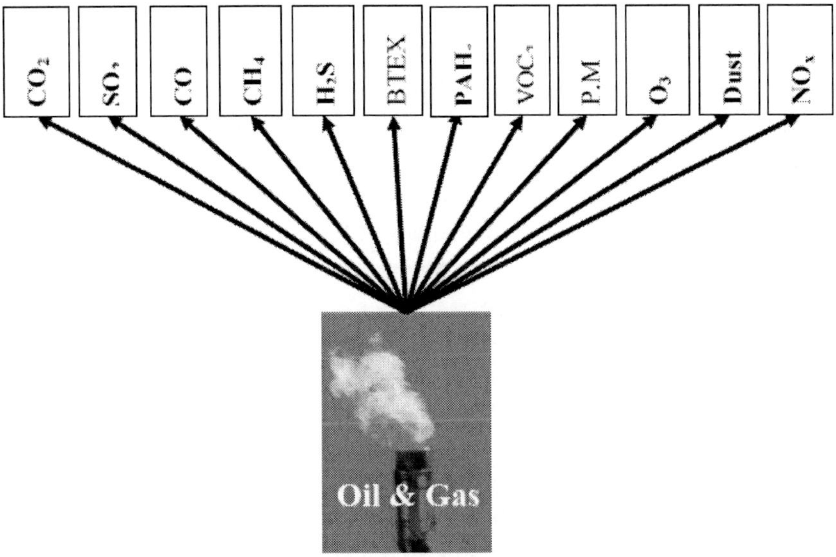

Figure 10. Schematic view of different pollution emission from oil and gas operations.

Project activities that would generate carbon emissions include vessel engines, generators and compressors on drill rigs and platforms, flaring, and other fugitive emissions such as spills.

Flaring

The petroleum industry carries out flaring, or the burning of waste gases, during well testing and in petroleum production operations (Figure 3). The main purposes of routine flaring are to dispose of unwanted or unusable volumes of gas and to depressure gas processing equipment for maintenance.

The current industry strategy for flaring is to monitor overall progress in reducing flaring; track progress in finding ideas that may reduce flaring; foster cooperation among industry to find innovative methods and initiate collective actions to reduce flaring; support research on flare combustion efficiency and alternative methods for using previously flared gases; and, and promote reassessment of regulatory mechanisms to remove impediments. In addition to this, the industry should promote a standardized monitoring scheme to assess how much carbon is released with every flare.

Engines

Engine efficiency is largely dependant on the way fuel is burned within the cylinders of the engine. The efficiency of the engines that are used for production and transportation should be strategically monitored in the oil and gas industry. In particular the amount of air emissions such as CO_2 that is released per unit fuel burned would be important data to keep. There are a number of CO_2 gas sensors on the market, which can monitor local area CO_2 content. For example the 4102 series, by Sierra Monitor Corporation is a group of self-contained, wall mounted CO_2 sensors, which detect local levels of CO_2 using the non-dispersive infrared method (SMC, 1996).

Spills

The environmental risk of an oil spill is defined as the product of (1) the probability of occurrence for the spill and (2) the environmental impact of the spill (Met Office, 2001). Relatively small spills occur frequently, but have low impact, whereas large spills occur rarely, but have high impact. Most contingency planning projects for specific areas or specific petroleum activities begin with an oil spill risk assessment. This means identification of the major sources of spills in the study area, and an assessment of the probability and impact of spills from these sources to identify "worst-case" and average-size spills that should be the focus of contingency planning and response preparedness. It is important to monitor whether or not a significant amount of carbon dioxide can evaporate into the atmosphere from a spill.

4.8. Other Projects / Activities and Cumulative Impacts

Cumulative impacts from developments occur when anthropogenically induced changes happen frequently in time and space and that the effects on an individual project cannot be assimilated (HAL and GMA, 2000). It is important to study the cumulative impacts of carbon because greenhouse gas emissions are one of the most important present-day issues. Besides Nova Scotia's oil and gas development projects, carbon is emitted from other past and present projects and activities occurring within or outside of the project area. Some of the past and present sources of carbon emissions include: domestic and international shipping, existing oil and gas exploration and production platforms, and engine emissions from vessels engaged in fishing, tourism, and military (naval) activities. There are many other sources, which are situated outside of the project area but will also add to the cumulative emissions of carbon (CNS, 2002). Some of the outside activities are: industrial development (especially from the United States), and road transportation.

In this study area, it is assumed the existing activities will continue to be carried out. The cumulative effects of operational emissions in combination with all existing projects/sources affecting the studying project (oil and gas development) and carbon levels might not be significantly increased from current levels but there is no doubt that the current emissions are significantly adding to the carbon budget.

CONCLUSIONS

In this study different emissions from oil and gas operations are identified and presented with respect to related activities. It is also revealed that oil and gas development in the study area, Nova Scotia, does not directly and indirectly impact on local carbon dioxide concentration. However, other pollutants release during operation, such as sulphur dioxide, particulate materials have local impacts and there is straight forward environmental assessment for this pollutants. This study also identified that considering the no/negligible impacts of carbon in the local area, the assessment of carbon is generally over-shadowed in most of environmental impact assessment studies. However, in this study, importance of carbon assessment is justified and a framework for the carbon assessment as a valued ecosystem is proposed. Analyzing different factors for carbon gives a greater scenario in local condition as well global conditions. For example, analysis of cumulative effects gives the past, present and future impacts of CO_2.

REFERENCES

Aunan, K. (2001) 'Climate measures in Norway: Potential reductions in offshore emissions', Internet Resource, <http://www.cicero.uio.no/edia/1265.pdf> [Accessed: 25 June 2004].

Berry, J.A. and Wells, G. (2004) 'Integrated Fate Modeling for Exposure Assessment of Produced Water on the Sable Island Bank (Scotian Shelf, Canada)', *Environmental Toxicology and Chemistry*, 23(10): 2483-2493.

Bjorndalen, N., Mustafiz, S., M.H. Rahman and Islam, M. R. (2005). 'No-Flare Design: Converting Waste to Value Addition', *Energy Sources*, Vol 27: 371-380.

Bott, R.D. (2006) *Sour Gas: Questions and Answers, Second Edition, Canadian Centre for Energy Information*, <www.centreforenergy. com>, ISBN 1- 894348-17-6.

Bouwman, A. F. (1989) 'Exchange of greenhouse gases between terrestrial ecosystems and the atmosphere'. *In Soils and the greenhouse effect.* A. F. Bouwman, ed. Pp 61 - 127. John Wiley and Sons, Chichester, UK. 575 pp.

Canada Nova Scotia Offshore Petroleum Board (CNSOPB), (2002) Environmental Protection Board. White Page. <http://www.cnsopb. ns.ca/Environment/evironment.html> (Accessed: April 21, 2002).

Canadian Association of Petroleum Producers (CAPP) (2002). *Technical Report. Produced Water Waste Management.* Canadian Association of Petroleum Producers, Canada, August 2002.

Canadian Association of Petroleum Producers (CAPP) (2006) *Environment, Protecting the Environment*, Canadian Association of Petroleum Producers, <http://www.capp.ca/>, [Accessed: March 10, 2007].

CBC, (2005) *Indepth: Environment Sour gas: Why it makes people nervous*, CBC News Online, August 10, 2005. <http://www.bc.ca/ews background/environment/
sour_gas.html>, [Accessed: March 10, 2007].

CNS (Clean Nova Scotia), (2002) *Energy (Coal, oil and gas, biomass, electricity and pipelines).*Contribution to Greenhouse Gas Emissions. <http://www.clean.ns.ca/> (Accessed: April 10, 2005).

Det Norske Veritas (DNV) *Environmental Consulting. 2000. Spill Modelling Tools. DNV. Norway.* <http://www.dnv.com/ environmentalcons/> [Accessed: 01 Apr 2002].

Earth Trend (2003) *Climate and Atmosphere—Canada*, <http://earthtrends.wri.org/pdf_library/country_profiles/cli_cou_124.pdf.> [Accessed: March 11, 2007].

EIP (Environmental Integrity Project) (2007) *Nine Refineries Spew 1/3 of Oil Industry Carcinogens*, Washington, DC, February 8, 2007 (ENS), <http://www.ens-newswire.com/ens/feb2007/2007-02-08-09.asp> [Accessed: March 09, 2007].

Energy Information Administration (EIA) (2002) *Norway: Environmental Issues*, Energy Information Administration <http://www.eia.doe.gov/ cabs/ norenv.html> (Cited: 25 April 2002).

Energy Information Administration (EIA) (2004) *International Energy Outlook 2004*, Greenhouse Gases general information, Energy Information Administration, Environmental Issues and World Energy Use. EI 30, 1000 Independence Avenue, SW, Washington, DC 20585.

Environment Canada (2006) *Informing Canadians on Air Pollution, Canada's National Pollutant Release Inventory*, <http://www.ec.gc.ca/ npri-inrp-comm/default.asp?lang=enandn=E786EDBA-1#star> [Accessed: March 10, 2007].

Esser, G. (1989) 'Modeling global terrestrial sources and sinks of CO_2 with special reference to soil organic matter'. In *Soils and the greenhouse effect*. A. F. Bouwman, ed. Pp 61 - 127. John Wiley and Sons, Chichester, UK. 575 pp.

Evoh, C. (2002) 'Gas Flares, Oil Companies and Politics in Nigeria, Chijoke Evoh is a graduate student at the Milano School of Public and Urban Policy', New School University, New York, USA, <http://www.waado. org/ Environment/OilCompanies/OilCompanies Page.html> [accessed: March 16, 2007].

GAIA. (2002). *Industrial Source Complex Dispersion Model,* GAIA. GAIA Environmental Information System <http://www.ess.co.at/ GAIA/models/ isc.htm> (Cited:29 Mar 2002).

Government of Nova Scotia (2001) *Nova Scotia Energy Policy 2001*, Government of Nova Scotia, Halifax, Canada.

Haites. E, (2000) *Possible domestic policies to manage greenhouse gas emissions*, Agriculture and Agri-food Canada, Publication no. 2034.

HAL and GMA. (2000) *Workshop on cumulative environmental affects assessment and monitoring on the Grand Banks and Scotian Shelf*, (Hatch Associates Limited and Griffiths Muecke Associates), Environmental Studies Research Funds Report, ESRF 137, Ottawa, Ontario.

Haugan, P.M. (1998) 'Impacts on the Marine Environment from Direct and Indirect Ocean Storage of CO_2'. *Waste Management*, Vol. 17: 323-327.

Houghton, J. (2001) 'The science of global warming', *Interdisciplinary Science Reviews*. Vol. 25(4): 247-257.

Itsubo, N. and Yamamoto, R. (1999) 'Application of life cycle assessment to manufacturing of nonferrous metals', *Journal of the Japan Institute of Metals,* Vol. 63 (2): 208-214.

Kessel, D. G. (2000). 'Global warming - facts, assessment, countermeasures', *Journal of Petroleum Science and Engineering,* vol. 26(1-4): 157- 168.

Khan, M.I. (2006) *Towards Sustainability in Offshore Oil and Gas Operations*, Ph.D. Dissertation, Faculty of Engineering, Dalhousie University, Canada.

Khan, M.I. and Islam, M.R. (2005) 'Environmental modeling of oil discharges from produced water in the marine environment', *Proce. Of 33rd Annual General Conference of the Canadian Society for Civil Engineering 2005*,

Toronto, Ontario, Canada, June 2-4, 2005 / 2-4 juin 2005, Paper number: GC 225. 1-11 pp.

Khan, M.I. and Islam, M.R. (2007) *Handbook of Sustainable Oil and Gas Engineering Operations Management*, Gulf Publishing Company, Austin, Texas: 500 pp. [ISBN: 1-93376-212-8].

Kwon, O.Y. and Schnoor, J.L. (1994) 'Simple Global Carbon Model – The Atmosphere- Terrestrial Biosphere-Ocean Interaction', *Global Biogeochemical Cycles*, Vol. 8: 295-305.

Met Office (2001) *Ocean Carbon Cycle Modelling*. Met Office < http://www.met-office.gov.uk/research/ocean/climate/carbon.html> [Accessed: 21 Apr 2005).

National Climate Change Secretariat (1998) *Background Information on the Ability of the Industry to Contribute to Greenhouse Gas Emission Reductions*. National Climate Change Secretariat. Ottawa, Ontario.

Organization for Economic Cooperation and Development: International Energy Agency (OECD) (1991) *Greenhouse Gas Emissions: The Energy Dimension*. Organization for Economic Cooperation and Development: International Energy, Paris.

Patin, V. (1999) *Environmental Impact of the Offshore Oil and Gas Industry*, EcoMonitor Publishing: East Northport, New York.

Peterson, C. H., McDonald, L. L. and Green, R. H. (2001) 'Sampling design begets conclusions: the statistical basis for detection of injury to and recovery of shoreline communities after the 'Exxon Valdez' oil spill', *Marine Ecology – Progress Series*, Vol. 210: 255-283.

Raven, J.A. and Falkowski, P.G. (1999) 'Ocean Sinks for Atmospheric CO_2', *Plant Cell and Environment*, Vol. 22: 741-755.

Rotmans J. (1990) *IMAGE: An Integrated Model to Assess the Greenhouse Effect*. Kluwer Academic Publishers, Boston.

Salter, E. and Ford, J. (2001) 'Holistic environmental assessment and offshore oil field exploration and production' *Marine Pollution Bulletin*, Vol. 42 (1): 45-58.

Sierra Monitor Corporation (SMC). (1996) *Model 4102 Series: Carbon Dioxide Gas Sensor Modules. Sierra Monitor Corporation,* <http://www.sierramonitor.com/pdf/4102-XPDF> (Cited:01 Apr 2002).

UKOOA (United Kingdom Offshore Operations Association) (2002) *Atmospheric Emission. Available*, http://www.ukooa.co.uk/issues/1999report/ enviro99atmospheric.htm [March 11, 2007].

United Nations Environment Program (UNEP). (2002) *Offshore Oil and Gas Environment Forum*, <http://oef.unep.ch/emissions/ gasrel/ indehtm> [Accessed: 25 April 2006].

United Nations Framework Convention on Climate Change (UNFCC) (2006) *How human activities produce greenhouse gases*, Climate Change Information Sheet, <http://unfccc.int/essential_background/background_publications_htmlpdf/climate_change_information_kit/items/293.php>. [Accessed: March 19, 2007].

INDEX

A

B

C

D

E

F

G

H

I

J

K

L

M

N

O

P

T

U

V

W

Y

Z